Communications
in Computer and Information Science   949

Commenced Publication in 2007
Founding and Former Series Editors:
Phoebe Chen, Alfredo Cuzzocrea, Xiaoyong Du, Orhun Kara, Ting Liu,
Dominik Ślęzak, and Xiaokang Yang

More information about this series at http://www.springer.com/series/7899

Jian Chen · Yuji Yamada
Mina Ryoke · Xijin Tang (Eds.)

Knowledge and Systems Sciences

19th International Symposium, KSS 2018
Tokyo, Japan, November 25–27, 2018
Proceedings

Springer

Editors
Jian Chen
School of Economics and Management
Tsinghua University
Beijing, China

Mina Ryoke
Faculty of Business Sciences
University of Tsukuba
Tokyo, Japan

Yuji Yamada
Graduate School of Business Sciences
University of Tsukuba
Bunkyō, Tokyo, Tokyo, Japan

Xijin Tang
Institute of Systems Science
CAS Academy of Mathematics and Systems
Sciences
Beijing, China

ISSN 1865-0929   ISSN 1865-0937   (electronic)
Communications in Computer and Information Science
ISBN 978-981-13-3148-0   ISBN 978-981-13-3149-7   (eBook)
https://doi.org/10.1007/978-981-13-3149-7

Library of Congress Control Number: 2018960676

This Springer imprint is published by the registered company Springer Nature Singapore Pte Ltd.
The registered company address is: 152 Beach Road, #21-01/04 Gateway East, Singapore 189721, Singapore

Preface

The annual International Symposium on Knowledge and Systems Sciences aims to promote the exchange and interaction of knowledge across disciplines and borders to explore the new territories and new frontiers. With over 18-year continuous endeavors, attempts to strictly define knowledge science may be still ambitious, but a very tolerant, broad-based, and open-minded approach to the discipline can be taken. Knowledge science and systems science can complement and benefit each other methodologically.

The First International Symposium on Knowledge and Systems Sciences (KSS 2000) was initiated and organized by the Japan Advanced Institute of Science and Technology (JAIST) in September of 2000. Since then KSS 2001 (Dalian), KSS 2002 (Shanghai), KSS 2003 (Guangzhou), KSS 2004 (JAIST), KSS 2005 (Vienna), KSS 2006 (Beijing), KSS 2007 (JAIST), KSS 2008 (Guangzhou), KSS 2009 (Hong Kong), KSS 2010 (Xi'an), KSS 2011 (Hull), KSS 2012 (JAIST), KSS 2013 (Ningbo), KSS 2014 (Sapporo), KSS 2015 (Xi'an), KSS 2016 (Kobe), and KSS 2017 (Bangkok) have been a successful platform for many scientists and researchers from different countries. During the past 18 years, people interested in knowledge and systems sciences have become a community, and an international academic society has existed for 15 years.

This year KSS was held in Tokyo, Japan, during November 25–27, 2018. The conference provided opportunities for presenting interesting new research results and facilitating interdisciplinary discussions, leading to knowledge transfer under the theme of "Knowledge Acquisition from Structured and Unstructured Data for Effective Social Implementation." To fit that theme, four distinguished scholars delivered the keynote speeches.

- Chonghui Guo (Dalian University of Technology, China), "Big Data Analytics in Health Care: Data driven Methods for typical Diagnosis and Treatment Pattern Mining"
- Setsuya Kurahashi (University of Tsukuba, Japan), "Model-Based Policy Making: Urban Dynamics, Collaborative Learning, and Family Strategy"
- Yoichi Motomura (Artificial Intelligence Research Center, National Institute of Advanced Industrial Science and Technology, Japan), "Toward Cyber Physical Innovation: Probabilistic Modeling for Real-Field AI Applications"
- Thanaruk Theeramunkong (Sirindhorn International Institute of Technology, Thammasat University, Thailand), "Text Mining from Public Hearing Databases and Automatic Profile Generation from Online Resources"

KSS 2018 received 54 submissions from authors studying and working in Belgium, China, India, Indonesia, Japan, Sri Lanka, Thailand, and Russia, and finally 20 submissions were selected for publication in the proceedings after a double-blind review process. The co-chairs of the international Program Committee made the final decision for each submission based on the review reports from the referees, who came from China, Japan, New Zealand, Thailand, and the USA.

We received a lot of support and help from many people and organizations. We would like to express our sincere thanks to the authors for their remarkable contributions, all the Technical Program Committee members for their time and expertise in reviewing the papers within a very tight schedule, and the proceedings publisher Springer for their professional help. It is the third time that the KSS proceedings are published as a CCIS volume after successful collaboration with Springer in 2016 and 2017. We greatly appreciate our four distinguished scholars for accepting our invitation to deliver keynote speeches at the symposium. Last but not least, we are very indebted to the local organizers for their hard work.

We were happy to witness the thought-provoking and lively scientific exchanges in the fields of knowledge and systems sciences during the symposium.

November 2018   Jian Chen
                                                                    Yuji Yamada
                                                                    Mina Ryoke
                                                                     Xijin Tang

Organization

Organizer

International Society for Knowledge and Systems Sciences

Host

University of Tsukuba, Tokyo Campus, Japan

General Chairs

Jian Chen   Tsinghua University, China
Yuji Yamada   University of Tsukuba, Japan

Program Committee Chairs

Mina Ryoke   University of Tsukuba, Japan
Xijin Tang   CAS Academy of Mathematics and Systems Science,
                         China
Jiangning Wu   Dalian University of Technology, China

Technical Program Committee

Quan Bai   Auckland University of Technology, New Zealand
Masataka Ban   University of Tsukuba, Japan
Meng Cai   Xidian University, China
Zhigang Cao   Beijing Jiaotong University, China
Hao Chen   Nankai Univertsity, China
Jindong Chen   Beijing Information Science and Technology University,
                         China
Zengru Di   Beijing Normal University, China
Yong Fang   CAS Academy of Mathematics and Systems Science,
                         China
Chonghui Guo   Dalian University of Technology, China
Jun Huang   Angelo State University, USA
Van-Nam Huynh   Japan Advanced Institute of Science and Technology,
                         Japan
Setsuya Kurahashi   University of Tsukuba, Japan
Bin Jia   Beijing Jiaotong University, China
Jiradett Kerdsri   Defense Technology Institute, Thailand
Weidong Li   Shaanxi Normal University, China

Abstracts of Keynotes

Big Data Analytics in Healthcare: Data-Driven Methods for Typical Diagnosis and Treatment Pattern Mining

Chonghui Guo

Institute of Systems Engineering, Faculty of Management and Economics,
Dalian University of Technology, China
dlutguo@dlut.edu.cn

Abstract. A huge volume of digitized clinical data is generated and accumulated rapidly since the widespread adoption of Electronic Medical Records (EMRs). These massive quantities of data hold the promise of propelling healthcare evolving from a proficiency-based art to a data-driven science, from a reactive mode to a proactive mode, from one-size-fits-all medicine to personalized medicine. Personalized medicine refers to tailoring medical diagnosis and treatment to the individual characteristics of each patient, which literally means the ability to classify individuals into subpopulations that differ in their susceptibility to a disease or their response to a specific treatment. While EMRs contain rich temporal and heterogeneous medical information that can be used for typical diagnosis and treatment pattern mining by big data analytics. Hence, this study will analyze different data types of EMRs in depth and design data-driven EMRs mining method, including data-driven typical diagnosis pattern extraction from multi-type data of EMRs, and data-driven typical treatment pattern extraction from multi-view of doctor orders. Specifically, for typical diagnosis pattern extraction, we first design three similarity measure methods for patient demographic, symptom, and laboratory examination information, then adopt similarity fusion method to generate a unified similarity and construct similarity network of patient hospital admission, next propose a patient diagnostic information similarity method by integrating patient hospital admission information, and finally perform clustering algorithm to extract typical diagnosis patterns. For typical treatment pattern extraction, we first study automatic treatment regimen development and recommendation from the content view of doctor orders, then study typical treatment process extraction and evaluation from the sequence view of doctor orders, next study typical drug use pattern extraction and evaluation from the duration view of doctor orders, and finally propose a fusion framework for typical treatment pattern extraction from multi-view of doctor orders. Furthermore, all proposed methods have been validated on real-world EMRs of cerebral infarction dataset and MIMIC-III dataset.

Model-Based Policy Making: Urban Dynamics, Collaborative Learning and Family Strategy

Setsuya Kurahashi

Faculty of Business Sciences, University of Tsukuba, Japan
kurahashi.setsuya.gf@u.tsukuba.ac.jp

Abstract. Many significant policies of our society and economy are determined by someone day after day. However, most of the plans have been discussed and decided based on past experiences and data. Many of them estimate policy effects by analyzing actual phenomena and data using statistical methods. For this method called evidence-based policymaking (EBP), this lecture proposes model-based policymaking (MBP). The MBP is designed with an agent-based model and data science techniques, and it also called as social simulation. The model-based approach enables to design realistic phenomena as a model and predict the effect on unfolding future events due to hypotheses or activities that are difficult to experiment using computer experiments. In the field of business and sociology, data analysis as an induction method and strategy planning as a deductive method are connected. In the lecture, I will introduce urban dynamics model, teaching model at school, analysis of education in a family using a genealogy in China during 500 years.

Toward Cyber Physical Innovation: Probabilistic Modeling for Real Field AI Applications

Yoichi Motomura

Artificial Intelligence Research Center,
National Institute of Advanced Industrial Science and Technology, Japan
y.motomura@aist.go.jp

Abstract. Currently, the practical application of artificial intelligence is dramatically advanced by machine learning using real world big data. Industrial structure reform and the smart society called Society 5.0 are also expected to be realized. In this talk, real world application and the research projects on AI are introduced. Our social system is changing into a Cyber Physical System by AI and real world big data. For example, point of sales data (POS-data) is linked to the customer ID by the common point card system. Moreover, sensors in a store can capture a customer's behavior during shopping process. From phenomena of real world big data with high temporal and spatial resolution, phenomena can be represented by a probabilistic model that can be calculated, predicting risk, cost, benefit and making it possible to simulate. In our system, this computational process is realized by PLSA (Probabilistic Latent Semantic Analysis) and Bayesian networks.

We developed interactive digital signage systems and interactive vending machines driven by AI. These systems are being investigated in the use cases such as improvement of productivity of services such as management support and logistics optimization. The same framework can be utilized also child care and health promotion activities for local community support.

Text Mining from Public Hearing Databases and Automatic Profile Generation from Online Resources

Thanaruk Theeramunkong

Sirindhorn International Institute of Technology,
Thammasat University, Thailand
thanaruk@siit.tu.ac.th

Abstract. Analyzing natural language texts helps us to obtain information or knowledge for various purposes. This talk firstly provides a short summary of the state of the art on research and development in natural language processing, including language characteristics, rule-based and statistical methodologies, as well as difficulties and challenges. Secondly, with the growing availability of fact-oriented and/or opinion-rich online textual contents, I present new opportunities and challenges of using text mining techniques to seek out and understand the facts and the opinions in our society. Thirdly, along with this trend, electronic public hearing information, incident information and personal activity logs are described and the potentials towards knowledge discovery are enumerated and two applications of social monitoring and personal profile generation are discussed. For the first task, from 2014 to 2015, during the National Reform Council, there have been an activity of reform-related public hearing in Thailand. The information was applied for analyzing Thai opinion on the country's reform process and topics for reforming. The second task is to gather online information related to individual activities/events for generating personal profiles. Such information can be used for characterizing individuals for expert recruiting or seeking. In the talk, I report the progress of our research works on these two tasks, including their potential use in the future.

Keywords: Representation learning · Deep learning · COSFIRE trainable filters

Contents

Modeling the Heterogeneous Mental Accounting Impacts of Inter-shopping Duration

Kazuhiro Miyatsu[1]([⊠]) and Tadahiko Sato[2]

[1] The Nielsen Company, Tokyo, Japan
kazuhiro.miyatsu@nielsen.com
[2] Faculty of Business Sciences, University of Tsukuba, Tokyo, Japan
sato@gssm.otsuka.tsukuba.ac.jp

Abstract. Unlike the principles of traditional economics, namely that goods with monetary equivalency can be substituted, mental accounting states that these goods have different criteria values to consumers depending on the purposes of their use and circumstances at purchase. By modeling an inter-shopping duration that accommodates the mental condition changes captured by a newly formulated latent variable termed "mental loading" herein, our research examines how a consumer's mental factor affects his or her purchase behavior. From the perspective of behavioral economics, it models consumer purchase behaviors that are seemingly irrational from a traditional economics viewpoint. The model is derived from a threshold-based modeling framework that incorporates consumer heterogeneity in a hierarchical Bayesian manner, and the modeling parameters are estimated by using the Markov Chain Monte Carlo method. By using scanner panel data from a retailer, the empirical results show that our model outperforms those without consumers' mental condition changes at the time of purchase.

Keywords: Mental accounting · Inter-shopping duration
Threshold-based model

1 Introduction

In traditional economics, consumers are supposed to be rational in that they behave to maximize their utilities. However, consumers do not necessarily act as rational entities, and the principle of utility maximization may not always be applicable to describe all consumer behaviors. When goods are discounted, for example, consumers often buy them even if they are not in need. A rational economic entity would not risk wasting goods as it does not support utility maximization. To explain such seemingly irrational behavior, consumers' mental conditions can be considered. *Mental accounting* is a concept of behavioral economics that illustrates purchase behavior influenced by mental factors. In this research, we develop models of inter-shopping duration that take account of this mental accounting effect.

© Springer Nature Singapore Pte Ltd. 2018
J. Chen et al. (Eds.): KSS 2018, CCIS 949, pp. 1–16, 2018.
https://doi.org/10.1007/978-981-13-3149-7_1

Mental accounting was proposed by Thaler [1] as a concept of microeconomics combined with cognitive psychology. In contrast to assuming a monetary equivalency of goods as in traditional economics, mental accounting assumes that goods have different criteria values for consumers depending on the purpose of their use and circumstances at purchase. Thaler [2] recognized three types of heterogeneity in mental accounting: transaction utility, target category, and valuation frequency. To accommodate the heterogeneity in the models, we assume that consumers have two sets of response rates to the parameters in the likelihood function and that they change when a mental factor, associated with monthly cumulative spend, exceeds the threshold. In the mental accouting studies, new behavioral findings and their validity of the seemingly irrational rules were often presented in marketing and finacne, such as Marberis and Huang [3], Grinblatt and Han [4], Prelec and Loewenstein [5], Langer and Weber [6], and Shafir and Thaler [7]. But little has been studied in mental accounting to model at individual level.

As reciprocal to duration, Ehrenberg [8] proposed modeling the number of visits to a store within regular intervals, assuming a Poisson distribution with a heterogeneous rate in the Gamma distribution. A number of researchers have developed models based on Ehrenberg's framework, such as Charfield et al. [9], Morrison and Schmittlein [10], Sichel [11], and Gupta [12,13]. In marketing analytics, managers are most interested in knowing how media and promotion influence their customers' inter-shopping duration. Since Cox [14] presented a proportional hazard model, numerous studies have been conducted by using his framework. For instance, Seetharaman and Chintagunta [15] proposed a discrete hazard model that can be applied to discrete observations and Seetharaman [16] developed an additive hazard model to improve the estimation of the covariate effect.

We develop a hierarchical Bayesian model of inter-shopping duration, where two separate regimes exist for each household. Allenby et al. [17] proposed a hierarchical Bayesian model with three different regimes and applied for direct marketing of financial services. However, their estimating a transitional point between regimes is history data-driven, and it is not suitable for our model that incorporates structured parameters for separate regimes. Threshold-based approach is deployed in our model as is demonstrated by Ferreira [18], Geweke and Terui [19], Chen and Lee [20]. In marketing, Terui and Danaha [21] deployed the framework to estimate reference price regimes, and Terui and Ban [22] applied for ad-stock effect.

The remainder of this article consists of the following sections. Section 2 defines the models, and we derive the estimation algorithm in Sect. 3. In Sect. 4, the empirical study using a retailer's scanner panel data is presented, and Sect. 5 concludes.

2 Models

2.1 Mental Loading Model

In this study, a new concept to capture mental pressure at the time of purchase is introduced, called *mental loading*. This is defined as the cumulative spend at a point in time during two consecutive paydays. Upon being paid every month, mental loading is cleared and increases every time a new purchase is made. This reflects a consumer's mental condition at purchase; in other words, the consumer feels more pressure spending when mental loading is high. However, information on an individual's payday is not available for the models and differs by individual.

Let us first define *Loading_Period* in Eq. (1) to specify the three most common types of paydays. First, *loading_period* $l = 1$ corresponds to a payday on the 25th of every month, which would then continue until the 24th of the following month. In Japan, the three days in Eq. (1) are representative paydays:

$$Loading\_Period \begin{cases} l = 1 & \Rightarrow & 25^{th} \cdot N_{month} - 24^{th} \cdot (N+1)_{month} \\ l = 2 & \Rightarrow & 5^{th} \cdot N_{month} - 4^{th} \cdot (N+1)_{month} \\ l = 3 & \Rightarrow & 17^{th} \cdot N_{month} - 16^{th} \cdot (N+1)_{month} \end{cases} \quad (1)$$

Second, let $cumm_{i,t_i,l}$ be denoted as the cumulative spend for household i, purchase occasion at t_i, and loading_period of l in Eq. (2). For the first time after payday, $cumm_{i,t_i,l}$ is always 0 and purchase amount $M_{i,j}$ is added at every purchase until the next payday:

$$cumm_{i,t_i,l} = \begin{cases} \sum_{j=1}^{trans^l(t_i)-1}(M_{i,j}) & tran^l(t_i) \neq 1 \\ 0 & tran^l(t_i) = 1 \end{cases} \quad (2)$$

Finally, *mental loading* is defined as the linear combination of cumulative spend $cumm_{i,t_i,l}$ ($l = 1, 2, 3$) and weighting factor $\alpha_i^{*(k)}$ ($k = 1, 2, 3$) in Eq. (3), where $\alpha_i^{*(k)}$ has the constraints of $0 \leq \alpha_i^{*(k)} \leq 1$ and $\sum_{k=1}^{3} \alpha_i^{*(k)} = 1$, and $\alpha_i^* = (\alpha_i^{*(1)}, \alpha_i^{*(2)})$ are the mental loading structural parameters to estimate:

$$CummM_{i,t_i} = \alpha_i^{*(1)}cumm_{i,t_i,1} + \alpha_i^{*(2)}cumm_{i,t_i,2} + \alpha_i^{*(3)}cumm_{i,t_i,3} \quad (3)$$

Each household has its own structural parameters, which reflect the composition of the income earner(s) in a family. For example, in the case of $\alpha_i^* = (1, 0)$, a household has a single income earner paid on 25th of every month. When a household's parameter is $\alpha_i^* = (0.5, 0.5)$, this family should have two income earners with paydays on the 25th and 5th of every month. Figure 1 illustrates these two cases.

2.2 Inter-shopping Duration Model

Let us define inter-shopping duration as the number of days since the immediate previous occasion. For the shopping time t_{i,n_i} of household i on the nth

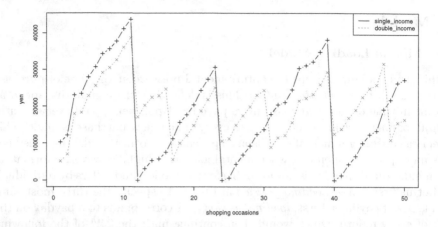

Fig. 1. Mental loading: single income and double income

occasion, the inter-shopping duration is calculated as $y_{i,t_i} = t_{i,n_i} - t_{i,n_i-1}$. For household i, our model has two shopping regimes depending on the degree of mental loading at time t_i compared with threshold $\gamma_{cum,i}$. The inter-shopping duration is assumed to be distributed log-normally for both regimes, whereas their distribution parameters differ. When mental loading $CummM_{i,t_i}$ at time t_i for household i exceeds threshold $\gamma_{cum,i}$, their shopping regime changes from Regime 2 to Regime 1. Equation (4) presents the inter-shopping duration model for household i at time t_{i,t_i}:

$$Pr(Y_{i,t_i} = y_{i,t_i} \mid \mu_i^{(1)}, \mu_i^{(2)}, \sigma_i^{2(1)}, \sigma_i^{2(2)}, \gamma_{cum,i}, CummM_{i,t_i}) \qquad (4)$$

$$= \begin{cases} \dfrac{1}{\sqrt{2\pi}\sigma_i^{(1)} y_{i,t_i}} \exp\left(\dfrac{-1}{2\sigma_i^{2(1)}} \{log(y_{i,t_i}) - \mu_i^{(1)}\}^2 \right), CummM_{i,t_i} \geq \gamma_{cum,i} \quad (Regime 1) \\[2ex] \dfrac{1}{\sqrt{2\pi}\sigma_i^{(2)} y_{i,t_i}} \exp\left(\dfrac{-1}{2\sigma_i^{2(2)}} \{log(y_{i,t_i}) - \mu_i^{(2)}\}^2 \right), CummM_{i,t_i} < \gamma_{cum,i} \quad (Regime 2) \end{cases}$$

Let location parameter $\mu_i^{(k)}(k = 1, 2)$ be assumed to have a linear combination structure with explanatory vector x_{i,t_i} for household i at time t_i; then, the parameters become time-dependent as the explanatory vector changes over time. Now, $\mu_i^{(k)}(k = 1, 2)$ is replaced with $\mu_{i,t_i}^{(k)} = x_{i,t_t}^t \beta_i^{(k)}(k = 1, 2)$, and the likelihood of Eq. (4) is multiplied for all shopping occasions until T_i, where T_i is the time of the last occasion of household i in the period. $R^{(k)}\{CummM_{i,t_i}, \gamma_{cum,i}\}$ is also introduced as an operator to determine to which regime it belongs. Assuming that all households are independent, the total likelihood is derived in Eq. (5):

$$p(\{y_{i,T_i}\} \mid \{\beta_i^{(1)}\}, \{\beta_i^{(2)}\}, \{\sigma_i^{2(1)}\}, \{\sigma_i^{2(2)}\}, \{\gamma_{cum,i}\}, \{CummM_{i,T_i}\}, \{x_{i,t_i}\}) \qquad (5)$$

$$= \prod_{i=1}^{H} \left\{ \prod_{k=1}^{2} \left\{ \prod_{t_i \in R^{(k)}\{CummM_{i,t_i}, \gamma_{inv,i}\}}^{T_i} \frac{1}{\sqrt{2\pi}\sigma_i^{(k)} y_{i,t_i}} \exp\left(\frac{-1}{2\sigma_i^{2(k)}} \{log(y_{i,t_i}) - x_{i,t_t}^t \beta_i^{(k)}{}^2\} \right) \right\} \right\}$$

2.3 Hierarchical Model

Because the model coefficients and structural parameters among households are assumed to have commonality in their attributes, a hierarchical structure is deployed as a linear function of these variables. By combining the regression coefficients in Eq. (5) and the threshold for each household, $\psi_i = (\beta_i^{(1)^t}, \beta_i^{(2)^t}, log(\gamma_{cum,i}))$ is defined. Likewise, mental loading weight vector $\alpha_i = (\alpha_i^{(1)}, \alpha_i^{(2)})$ is defined as an inverse logit function of $\alpha_i^{*(k)} = \dfrac{\exp(\alpha_i^{(k)})}{1 + \sum_{l=1}^{2} \exp(\alpha_i^{(l)})}$ $(k = 1, 2)$, and it possesses a similar hierarchical structure. Equations (6) and (7) are hierarchical models. For the variance parameters $\sigma_i^2 = (\sigma_i^{2(1)}, \sigma_i^{2(2)})$, a hierarchical mechanism is not constructed as a regression model in this study (we assume the prior distribution for these parameters):

$$\psi_i = z_i^t \Delta_\psi + \epsilon_i^\psi, \ \epsilon_i^\psi \sim \mathcal{N}(0, \Sigma_\psi) \tag{6}$$

$$\alpha_i = z_i^t \Delta_\alpha + \epsilon_i^\alpha, \ \epsilon_i^\alpha \sim \mathcal{N}(0, \Sigma_\alpha) \tag{7}$$

ψ_i and α_i are expressed as linear functions of household attribute vector $z_i = (z_{i,1}, z_{i,2}, ..., z_{i,q})^t$ and coefficient matrix Δ_ψ, Δ_α, where q is a dimension of the attribute vector. Σ_ψ and Σ_α are the variance-covariance matrixes of ψ_i and α_i, respectively.

For prior distributions, these matrixes are assumed to have a natural conjugate relationship of $p(\Delta_\psi, \Sigma_\psi) = p(\Delta_\psi | \Sigma_\psi)p(\Sigma_\psi), p(\Delta_\alpha, \Sigma_\alpha) = p(\Delta_\alpha | \Sigma_\alpha)p(\Sigma_\alpha)$, where $p(\Delta_\psi | \Sigma_\psi), p(\Delta_\alpha | \Sigma_\alpha)$ are multi-variable normal distributions and $p(\Sigma_\psi), p(\Sigma_\alpha)$ are inverse-Wishart distributions. As a result, the coefficients of the linear functions of the hierarchical structure are expressed in the following equations:

$$\delta_\psi = vec(\Delta_\psi) \mid \Sigma_\psi \sim \mathcal{N}(\bar{\delta}_\psi, \Sigma_\psi \otimes A_\psi^{-1}), \ \ \Sigma_\psi \sim \mathcal{IW}(\upsilon_{\psi,0}, V_{\psi,0}) \tag{8}$$

$$\delta_\alpha = vec(\Delta_\alpha) \mid \Sigma_\alpha \sim \mathcal{N}(\bar{\delta}_\alpha, \Sigma_\alpha \otimes A_\alpha^{-1}), \ \ \Sigma_\alpha \sim \mathcal{IW}(\upsilon_{\alpha,0}, V_{\alpha,0}) \tag{9}$$

3 Algorithm of the Model Estimation

Based on the models developed in the previous section, Fig. 2 depicts the directed acyclic graph of our proposed model. Equation (10) shows the decomposition of the simultaneous posterior distribution of the parameters. In the derivation of Eq. (10), we assume independence between households:

$$p(\{\psi_i\}, \{\alpha_i\}, \{\sigma_i^{2(1)}\}, \{\sigma_i^{2(2)}\} \mid \{y_{i,T_i}\}, \{x_{i,T_i}\}, \{cumm_{i,T_i}\}, \{z_i\})$$

$$\propto p(\Delta_\psi | \Sigma_\psi)p(\Sigma_\psi)p(\Delta_\alpha | \Sigma_\alpha)p(\Sigma_\alpha) \times$$

$$\prod_{i=1}^{H} \Big\{ p(\psi_i | \Delta_\psi, \Sigma_\psi, z_i)p(\alpha_i | \Delta_\alpha, \Sigma_\alpha, z_i)p(\sigma_i^{2(1)})p(\sigma_i^{2(2)}) \times$$

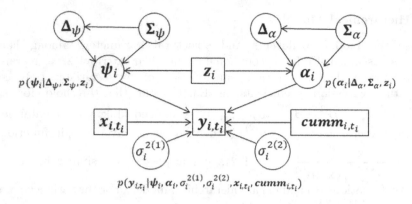

Fig. 2. DAG (Directed Acyclic Graph).

$$\prod_{k=1}^{2}\left\{\prod_{t_i \in R^{(k)}}{}_{\{cumm_{i,t_i},\gamma_{cum,i},\alpha_i\}} \frac{1}{\sqrt{2\pi}\sigma_i^{(k)}y_{i,t_i}} \exp\left(\frac{-1}{2\sigma_i^{2(k)}}\{log(y_{i,t_i}) - \beta_i^{(k)t}x_{i,t_i}^{(k)}\}^2\right)\right\}\right\}$$

(10)

The parameters included in the model are $\{\psi_i\}, \{\alpha_i\}, \{\sigma_i^{2(1)}\}, \{\sigma_i^{2(2)}\}$, $\Delta_\psi, \Delta_\alpha, \Sigma_\psi, \Sigma_\alpha$, among which $\{\psi_i\}, \{\alpha_i\}, \{\sigma_i^{2(1)}\}, \{\sigma_i^{2(2)}\}$ are heterogeneous to the household, while $\Delta_\psi, \Delta_\alpha, \Sigma_\psi, \Sigma_\alpha$ are common in all households. The Markov Chain Monte Carlo method is applied to estimate the parameters. Because $\Delta_\psi, \Delta_\alpha, \Sigma_\psi, \Sigma_\alpha$ are structured in natural conjugate relationships defined in Eqs. (8) and (9), these parameters can be drawn by using Gibbs samplers. On the contrary, $\{\psi_i\}, \{\alpha_i\}$ cannot formulate the conjugate structure, and thus it needs random walk samplers of the Metropolis-Hastings algorithm. $\sigma_i^{2(1)}, \sigma_i^{2(2)}$ are sampled independently by the Gibbs samplers:

$$\delta_\psi \mid \Sigma_\psi \sim \mathcal{N}(\tilde{\delta}_\psi, \Sigma_\psi \otimes (Z^t Z + A_\psi)^{-1}) \tag{11}$$

$$\Sigma_\psi \sim \mathcal{IW}(v_{\psi,0} + H, V_{\psi,0} + S_\psi) \tag{12}$$

where $\tilde{\delta}_\psi = vec(\tilde{\Delta}_\psi), \tilde{\Delta}_\psi = (Z^t Z + A_\psi)^{-1}(Z^t Z \hat{\Delta}_\psi + A_\psi \bar{\Delta}_\psi)$, $\hat{\Delta}_\psi = (Z^t Z)^{-1} Z^t \Psi$, $S_\psi = \sum_{i=1}^{H}(\psi_i - \bar{\psi}_i)(\psi_i - \bar{\psi}_i)^t$, $\Psi = (\psi_1, \psi_2, \ldots, \psi_H)$, $\bar{\delta}_\psi = 0$, $A_\psi = 10^{-6}I$, $v_{\psi,0} = 16$, $V_{\psi,0} = v_{\psi,0}I$

$$\delta_\alpha \mid \Sigma_\alpha \sim \mathcal{N}(\tilde{\delta}_\alpha, \Sigma_\alpha \otimes (Z^t Z + A_\alpha)^{-1}) \tag{13}$$

$$\Sigma_\alpha \sim \mathcal{IW}(v_{\alpha,0} + H, V_{\alpha,0} + S_\alpha) \tag{14}$$

where $\tilde{\delta}_\alpha = vec(\tilde{\Delta}_\alpha), \tilde{\Delta}_\alpha = (Z^t Z + A_\alpha)^{-1}(Z^t Z \hat{\Delta}_\alpha + A_\alpha \bar{\Delta}_\alpha)$, $\hat{\Delta}_\alpha = (Z^t Z)^{-1} Z^t \mathcal{A}$, $S_\alpha = \sum_{i=1}^{H}(\alpha_i - \bar{\alpha}_i)(\alpha_i - \bar{\alpha}_i)^t$, $\mathcal{A} = (\alpha_1, \alpha_2, \ldots, \alpha_H)$, $\bar{\delta}_\alpha = 0$, $A_\alpha = 10^{-6}I$, $v_{\alpha,0} = 16$, $V_{\alpha,0} = v_{\alpha,0}I$

When the likelihood function is log-normal, where $log(y_{i,t_i})$ is a normal distribution and the variance of the prior distribution is inverse-Gamma, the variance of posterior distribution $\sigma_i^{2(k)}(k = 1, 2)$ in Eq. (10) is also full-conditionally expressed in inverse-Gamma form with $\boldsymbol{\beta}_i^{(k)}$ in Eq. (15):

$$\sigma_i^{2(k)} \mid \boldsymbol{\beta}_i^{(k)} \sim IG(v_{i,n}^*/2, s_{i,n}^*/2), \quad k = 1, 2 \tag{15}$$

where $v_{i,n}^* v_0 + n_i$, $s_{i,n}^* = s_0 + (log(\boldsymbol{y}_i) - \boldsymbol{x}_i{}^t \boldsymbol{\beta}_i^{(k)})^t (log(\boldsymbol{y}_i) - \boldsymbol{x}_i{}^t \boldsymbol{\beta}_i^{(k)})$, $v_0 = 100$, $s_0 = 100$.

For the random walk sampling of $\boldsymbol{\psi}_i$, $\boldsymbol{\alpha}_i$, candidate samples are iteratively generated by $\boldsymbol{\psi}_i = \boldsymbol{\psi}_i^{(r-1)} + \delta_{RW_\psi}^2 \boldsymbol{I}$ and $\boldsymbol{\alpha}_i = \boldsymbol{\alpha}_i^{(r-1)} + \delta_{RW_\alpha}^2 \boldsymbol{I}$, where the variance in random walk noise is set as $\sigma_{RW_\psi}^2 = 10^{-5}, \sigma_{RW_\alpha}^2 = 10^{-2}$, respectively.

4 Empirical Analysis

4.1 Data

We estimate the models by using scanner panel data from a retailer's store in Tokyo between January 1 and December 31, 2001. Whenever a household member visits the store and makes a purchase, the details of the transaction are recorded for each household. Altogether, 100 sample panels were randomly selected of those who shopped more than 50 occasions and at least once every month in that period. Table 1 shows the mean and standard deviation (SD) of the variables and household attributes.

The explanatory variables consist of two types: household-specific and common for all households. The discount rate and items on the flyer are identical for all households at the time of shopping, whereas the number of items purchased on the previous shopping occasion differs by household. Unlike the brand choice model in marketing science, which concerns particular products or categories, we consider all the products in the store. The discount rate is defined as the average rate of products available at all times, which is 16% in the study period. The products examined in this study only include non-perishable food and sundries. Perishable food is merchandised on market quotations, and no discount rate is available. The number of items on the flyer includes perishable food, and the average amount is 37.2. As each household has its own family structure and shopping pattern, the number of items purchased per store visit differs. The average number of items per shopping is 14.3, of which 66% are non-perishable items. Household attributes are calculated based on holdout samples before the study period. Average spend per shopping trip is 1,537 JPY, and 61% of households are located within walking distance of the store. The region around the retail store has 55% of part-time workers.

4.2 Model Valuation

By using scanner panel data from the 100 randomly selected household panels, we estimated the heterogeneous parameters of the proposed model (Asymmetric Model) together with the two other models: the model with no regime (Null

Table 1. Definition of the variables and descriptive statistics

	Variable	Description	Mean	SD
y_{i,t_i}	$DRTN_{i,t_i}$	Inter-shopping duration	2.02	1.98
x_{i,t_i}	$CNST_{i,t_i}$	constant:1	1.00	0.00
	$PURCH_{i,t_i}$	The number of items purchased on the previous occasion	14.30	8.16
	$ITEMR_{i,t_i}$	The ratio of non-perishable items purchased on the previous occasion	0.66	0.21
	$DSCNT_{i,t_i}$	The average discount rate of non-perishable items on the current occasion	0.16	0.11
	$INSTM_{i,t_i}$	The square root of the number of flyers distributed on the current occasion	6.10	4.58
	$CummM_{i,t_i}$	*** Latent variable applicable only for Regime 1 ***	–	–
z_i	$LPAID_i$	log (the average payment amount per shopping occasion/1,000)	0.43	0.17
	$WORK1_i$	Work status type dummy 1 (home-maker = 1, others = 0)	0.30	0.46
	$WORK2_i$	Work status type dummy 2 (part-time = 1, others = 0)	0.55	0.50
	$WORK3_i$	Work status type dummy 3 (full-time = 1, others = 0)	0.11	0.31
	$FOOT_i$	Transportation dummy (on foot = 1, others = 0)	0.61	0.49

Table 2. Model valuation

	Null model	Symmetric model	Asymmetric model
Deviance information criteria [23] DIC	52,152.26	44,374.50	39,813.53
Log of the marginal likelihood [24] LML	−24,529.15	−22,845.47	−19,822.21

Model) and the threshold model with symmetric data without the mental loading variable (Symmetric Model). The Markov Chain Monte Carlo method was applied for 50,000 iterations; 45,000 samples were discarded as the burn-in period and the remaining 5,000 samples were used in the analysis. Table 2 summarizes the statistics for the model valuation. Both statistics support our proposed model (Asymmetric Model) following the Symmetric Model and Null Model.

4.3   Regression Parameter ($\beta_i^{(1)}$, $\beta_i^{(2)}$)

Table 3 shows the statistics for the average coefficient parameters of all the households in each regime. Under Bayesian statistics, heterogeneous coefficients are estimated for each household to analyze the impact on the inter-shopping duration by household. However, in this subsection, we discuss shopping behavior based on the average coefficients of all household panels. The intercept, which is a baseline of location parameter $\mu^{(k)}$, is $\beta_1^{(1)} > \beta_1^{(2)}$. Eliminating all the factors, the inter-shopping duration tends to be longer in Regime 1. However, with a negative sign of $\beta_6^{(1)}$, mental loading shortens this duration when it exceeds the threshold. As a result, the inter-shopping duration tends to be shorter in Regime 1. Regarding $\beta_2^{(k)}$, the more items a household purchases, the longer is the duration to the next shopping occasion. When a household possesses more goods in stock, it is rational to refrain from shopping, as there is no immediate need to visit the store. However, this is no longer the case or a household seems to be indifferent to the inter-shopping duration when it feels more pressure to spend. This phenomenon is noteworthy when more non-perishable goods are procured. For the ratio of non-perishable items, $\beta_3^{(1)} < \beta_3^{(2)}$ means that it becomes more insensitive when the regime switches to Regime 1. On the contrary, $\beta_4^{(1)} < \beta_4^{(2)}$ implies that the discount is more sensitive in Regime 1. Lastly, the expected sign of $\beta_5^{(k)}$ is negative, suggesting that the duration shortens when more items are on a flyer; however, this is only the case in Regime 2. Flyers are not as effective in Regime 1. In general, the inter-shopping duration becomes shorter in Regime 1, where a household feels more pressure to spend. From a purchase behavior viewpoint, households thus turn out to be more planned in Regime 1 and they visit a store to fulfill their immediate needs at any time.

4.4 Threshold Parameter ($\gamma_{cum,i}$)

Table 3 shows the statistics of the average threshold parameters of all the households. The average threshold is 11,649 JPY, indicating that regime switching occurs when mental loading exceeds 27.3% of monthly expenditure on average.

Understanding the relationship between $\gamma_{cum,i}$ and $\beta_{6,i}^{(1)}$ would help explain how mental loading affects the duration. Figure 3 shows the scatter plots of $\gamma_{cum,i}$ and $\beta_{6,i}^{(1)}$. The higher the threshold, the less effective is the response of mental loading. Households with a lower threshold perceive pressure to spend at an earlier stage, raising the impact of mental loading. As a result, the inter-shopping duration shortens and thus a household purchases goods more frequently to fulfill their immediate needs.

4.5 Mental Loading Structural Parameter ($\alpha_i^{(1)}$, $\alpha_i^{(2)}$)

Mental loading is the latent variable, and the structural parameters determine the composition of income earners based on their monthly payday dates. Figure 4 shows the scatter plots of $(\alpha_i^{*(1)}, \alpha_i^{*(2)})$. The plots on $(1,0), (0,1), (0,0)$ correspond to single-income households with a loading period $l = 1, 2, 3$. The plots on the $\alpha_i^{*(1)} + \alpha_i^{*(2)} = 1$, $\alpha_i^{*(2)}$, and $\alpha_i^{*(1)}$ axes indicate double-income households with the loading periods $l = (1,2), (2,3), (3,1)$, respectively; the others are from

Table 3. Basic statistics of the estimated parameters

Coefficient	Mean	Median	Min	Max	SD	Rate*
$\beta_{1,i}^{(1)}$	0.4052	0.3900	−0.7397	1.4549	0.3760	82%
$\beta_{2,i}^{(1)}$	−0.0068	0.0036	−0.5043	0.1586	0.0789	22%
$\beta_{3,i}^{(1)}$	0.1499	0.1351	−0.8320	0.6743	0.2381	65%
$\beta_{4,i}^{(1)}$	0.0577	0.0509	−0.7608	1.4232	0.4127	70%
$\beta_{5,i}^{(1)}$	0.0142	0.0007	−0.2914	0.9730	0.1101	19%
$\beta_{6,i}^{(1)}$	−0.0453	−0.0598	−0.3933	0.6258	0.1554	66%
$\beta_{1,i}^{(2)}$	0.3724	0.3298	−0.6494	1.5750	0.4618	84%
$\beta_{2,i}^{(2)}$	0.0502	0.0248	−0.3706	0.8450	0.1500	39%
$\beta_{3,i}^{(2)}$	0.2434	0.2416	−1.2171	1.1805	0.4380	76%
$\beta_{4,i}^{(2)}$	0.1151	0.1983	−0.9531	1.0491	0.4491	81%
$\beta_{5,i}^{(2)}$	−0.0166	0.0012	−0.5850	0.5322	0.1557	31%
$\sigma_i^{2(1)}$	0.5962	0.5631	0.2455	1.0129	0.2160	81%
$\sigma_i^{2(2)}$	0.9263	0.9649	0.5981	1.0231	0.0949	10%
$\gamma_{cum,i}$	11,649	7,554	440	111,679	14,887	100%
$\alpha_i^{(1)}$	0.3333	0.2600	0.0000	1.0000	0.3671	63%
$\alpha_i^{(2)}$	0.3341	0.3393	0.0000	1.0000	0.3464	67%

Rate: 95% statistical significance of all households based on the quasi t-value (Mean/SD)

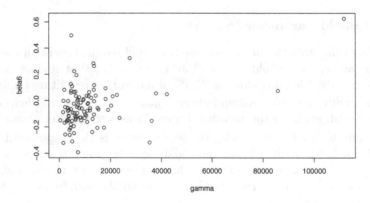

Fig. 3. Scatter plots ($\gamma_{cum} - \beta_6$)

triple- income households. Figure 4 also summarizes the occupancy rate of the composition of income earners among all household panels, counted based on estimating α_i^* by disregarding values below 0.01. The shares of single-, double-, and triple- income households are 39%, 42%, and 19%, respectively. In most companies in Japan, employees are paid every month; however, the payday date differs largely depending on the size and ownership status of the company: Table 3 shows the statistics of the average threshold parameters of all the households. The average threshold is 11,649 JPY, indicating that regime switching occurs when mental loading exceeds 27.3% of monthly expenditure on average.

Understanding the relationship between $\gamma_{cum,i}$ and $\beta_{6,i}^{(1)}$ would help explain how mental loading affects the duration. Figure 3 shows the scatter plots of $\gamma_{cum,i}$ and $\beta_{6,i}^{(1)}$. The higher the threshold, the less effective is the response of mental loading. Households with a lower threshold perceive pressure to spend at an earlier stage, raising the impact of mental loading. As a result, the inter-shopping duration shortens and thus a household purchases goods more frequently to fulfill their immediate needs. $l = 1$ is applicable for large publicly listed corporations, $l = 2$ for private and small or medium-sized companies, and $l = 3$ for government and public service entities. At a glance, the proportion of triple-income households seems higher. As family size has deceased in Japan, the triple- income earners in a family are higher than expected. However, most part-timers have two paydays a month, so this is treated as different income earners in this estimation.

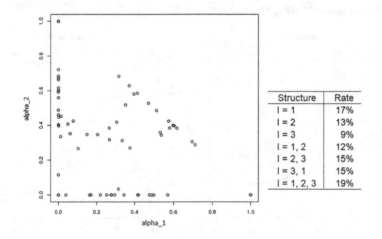

Fig. 4. Mental loading structural vector

4.6   Hierarchical Model Parameter $(\delta_1, \delta_2, \delta_3, \delta_4, \delta_5)$

The hierarchical model parameters reveal the common impact on the different coefficients. Table 4 shows the estimated results for the hierarchical model parameters. No obvious commonalities are identified in most cases, which is partly because of the insufficient number of household panels in the model, except for γ_{cum}. The threshold parameter is estimated in a stable manner, and the levels of significance of the attribute variables are high except for $FOOT$. The degree of average spend $LPAID$ influences $\beta_1^{(1)}$ and $\beta_1^{(2)}$. This average spend has a negative impact on the intercept in Regime 1, while it is positive in Regime 2. Part-time workers have a positive impact on $\beta_3^{(2)}$ and $\beta_4^{(2)}$ only in Regime 2.

Table 4. Hierarchical model estimation

VariableＡ@	CoefficientＡ@	$LPAID$ δ_1	$WORK1$ δ_2	$WORK2$ δ_3	$WORK3$ δ_4	$FOOT$ δ_5
$CNST^{(1)}$	$\beta_1^{(1)}$	* -0.3789	***0.4862	***0.5501	***0.5762	0.0630
$CNST^{(2)}$	$\beta_1^{(2)}$	*0.4888	0.2409	0.0841	-0.0586	0.0432
$PURCH^{(1)}$	$\beta_2^{(1)}$	-0.0177	0.0071	-0.0101	0.0127	0.0038
$PURCH^{(2)}$	$\beta_2^{(2)}$	-0.0229	-0.0210	0.0232	-0.0229	0.0793
$ITEMR^{(1)}$	$\beta_3^{(1)}$	0.1583	0.1222	-0.0176	0.0267	0.0859
$ITEMR^{(2)}$	$\beta_3^{(2)}$	0.1366	0.0840	0.1425	**0.3939	0.0487
$DSCNT^{(1)}$	$\beta_4^{(1)}$	-0.2884	0.0893	0.0677	0.0996	**0.1530
$DSCNT^{(2)}$	$\beta_4^{(2)}$	-0.2109	0.1492	0.1219	***0.5140	0.0617
$INSTM^{(1)}$	$\beta_5^{(1)}$	0.0202	0.0160	0.0294	0.0086	-0.0266
$INSTM^{(2)}$	$\beta_5^{(2)}$	-0.0052	0.0651	-0.0035	0.0372	-0.0535
$CummM^{(1)}$	$\beta_6^{(1)}$	*0.3398	*-0.1828	*-0.1709	*-0.2140	-0.0261
$THRSH$	γ_{cum}	***6.5761	***6.0261	***5.8488	***5.9571	0.4514
$ALPH^{(1)}$	$\alpha^{(1)}$	**4.6526	** -36.8833	-17.3374	** -43.1437	-4.6015
$ALPH^{(2)}$	$\alpha^{(2)}$	18.8939	-20.6767	-16.7853	-19.8601	1.9962

Statistical significance: ***95%, **90%, *80%.

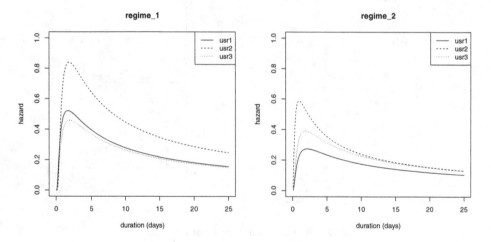

Fig. 5. Hazard (a Double-Regime Occupant)

4.7   Inter-shopping Duration Hazard

The parametric hazard function is derived based on the distribution parameters of the log-normal probability density function. Although variance is a fixed parameter for each household, the location parameter is designed to have a linear structure with the explanatory variables for household i at time t_i. Consequently, the location parameter possesses time dependency, meaning that the hazard function also changes on every occasion. In this subsection, we discuss the hazard functions for the households listed in Table 5, where the location

Table 5. Regime occupancy and durations

	Location		Variance		Occupancy		Duration		Threshold
	R1	R2	R1	R2	R1	R2	R1	R2	$\gamma_{cum,i}$
usr1	0.2525	2.1217	0.4109	1.0128	99%	1%	1.44	8.00	3,744
usr2	0.4178	1.3518	0.5095	0.9744	93%	7%	1.71	5.62	18,056
usr3	0.7312	1.0701	0.5095	0.8622	26%	74%	2.34	3.69	14,056
usr4	0.1038	–	0.3005	–	100%	0%	1.17	–	3,269
usr5	0.3885	–	0.4113	–	100%	0%	1.67	–	4,985
usr6	0.3584	–	0.4335	–	100%	0%	1.59	–	8,366
usr7	–	0.7772	–	0.6464	0%	100%	–	2.62	3,744
usr8	–	0.8376	–	0.8382	0%	100%	–	3.34	18,056
usr9	–	1.0801	–	0.6565	0%	100%	–	3.39	14,810

R1: $CummM_{i,t_i} \geq \gamma_{cum,i}$ $(Regime1)$, R2: $CummM_{i,t_i} < \gamma_{cum,i}$ $(Regime2)$
Location : $(1/N_i)\sum_{n=1}^{N_i} \mu_{i,n}^{(k)}$, Duration : $(1/N_i)\sum_{n=1}^{N_i} y_{i,n}^{(k)}, (k = 1, 2)$

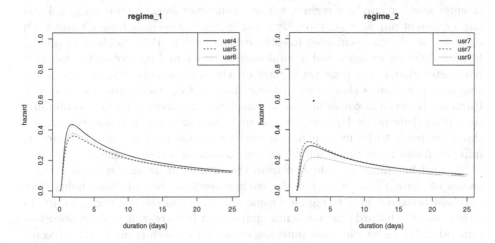

Fig. 6. Hazard (a Single-Regime Occupant)

parameters are calculated by using the average values of the explanatory vector and duration is the average inter-shopping duration for each regime.

As usr1–3 have shopping occasions in each regime, they have different hazard functions, which are not monotonously increasing or decreasing but unimodal. The location and height of a peak are different for each household. Figure 5 illustrates the hazard functions of the users. They all have sharp-pointed peaks in Regime 1. Their purchase behavior changes once they enter Regime 1, and they shift the mindset of paying attention to fulfill their immediate needs; thus, the inter-shopping duration becomes shorter. Usr4–6 and usr7–9 have shopping occasions in a single regime only. Figure 6 show the hazard functions of usr4–6 on the left and usr7–9 on the right. Whenever a single regime remains, their hazard functions resemble each other, and the peaks are not too high. Usr4–6 should perceive pressure to spend from the initial shopping occasion, but their hazard functions do not have sharp-pointed peaks. The average duration in Regime 1 is already short, and their purchase behavior always aims to fulfill their immediate needs. On the contrary, usr7–9 never move into Regime 1 and they should not receive much pressure to spend, meaning that the average duration is relatively longer. Although all the hazard functions in Fig. 6 seem alike, the purchase attitudes of usr4–6 and usr7–9 are quite different.

5   Conclusion

In this study, we developed models of inter-shopping duration based on the mental accounting effect. As a proxy of the consumer's mental index at the time of purchase, *mental loading* was introduced to capture changes in their mental conditions. Mental loading was defined as the cumulative spend from one monthly payday to the next. When this exceeds the threshold, it is assumed to enter another purchase regime, where a consumer feels pressure to spend and has a different purchase attitude. By using scanner panel data from a retailer, an empirical study was conducted to estimate the models. Households were proved to have different regimes, and mental loading was found to explain the location parameter that has a negative impact on the duration. In general, the inter-shopping duration is shorter when the consumer feels more pressure to spend. Purchase behavior becomes myopic as consumers pay more attention to fulfilling their immediate needs. By contrast, feature is no longer as effective as consumers' shopping tends to be more planned. Further, goods in stock are no longer as influential and those goods purchased are consumed rather quickly.

Hazard functions were derived from the estimated parameters and average values of the explanatory vector for each household. Not all households have purchase occasions in both regimes; some stay in one of the two regimes only. In either case, the hazard function is not monotonously increasing or decreasing but unimodal. For those who have purchase occasions in both regimes, their hazard functions have abrupt high peaks when mental loading exceeds the threshold. On the contrary, for those whose purchase occasions are in a single regime only, the peaks of their hazard functions are gently sloping. As shown in Table 5, regardless of the regime, the inter-shopping duration is shorter when the consumer feels pressure to spend. Abrupt peaks appear only when the household changes regime. Indeed, changes in mental conditions lead to behavioral changes in shopping. Even if consumers are exposed to an environment characterized by high pressure to spend, the instantaneous probability of shopping would not be affected unless changes in mental conditions occur.

We demonstrated that the deployment of soft factors into hard facts such as scanner panel data in modeling can help understand consumers from a behavioral economics viewpoint. Mental accounting is one of such robust soft factors that we proved to successfully incorporated into the models by introducing the notion of mental loading and its threshold.

Acknowledgements. This work was supported by JSPS Grand-in-Aid for Scientific Research (B) Grant Number JP18H00904.

References

1. Thaler, R.: Mental accounting and consumer choice. Mark. Sci. **4**, 199–214 (1985)
2. Thaler, R.: Mental accounting matters. Mark. J. Behav. Decis. Mak. **12**, 183–206 (1999)
3. Marberis, N., Huang, M.: Mental accounting, loss aversion, and individual stock returns. J. Financ. **56**, 1247–1292 (2001)
4. Grinblatt, M., Han, B.: Prospect theory, mental accounting, and momentum. J. Financ. Econ. **78**, 311–339 (2005)
5. Prelec, D., Loewenstein, G.: The red and the black: mental accounting of savings and debt. Mark. Sci. **17**, 4–28 (1998)
6. Langer, T., Weber, M.: Prospect theory, mental accounting, and differences in aggregated and segregated evaluation of lottery portfolios. Manag. Sci. **47**, 716–733 (2001)
7. Shafir, E., Thaler, R.: Invest now, drink later, spend never: on the mental accounting of delayed consumption. J. Econ. Psychol. **27**, 694–712 (2006)
8. Ehrenberg, A.S.: The pattern of consumer purchases. J. R. Stat. Soc. **8**, 26–41 (1959)
9. Charfield, C., Ehrenberg, A.S., Goodhar, G.J.: Progress on a simplified model of stationary purchasing behavior. J. R. Stat. Soc. **68**, 828–835 (1966)
10. Morrison, R.G., Schmittlein, D.C.: Generalizing the NBD model for customer purchases: what are the implication and is it worth the effort? J. Bus. Stat. **6**, 145–159 (1988)
11. Sichel, P.E.: Repeat-buying and the generalized inverse Gaussian-Poisson distribution. J. R. Stat. Soc. **31**, 193–204 (1982)
12. Gupta, S.: Stochastic models of interpurchase time with time-dependent covariates. J. Mark. Res. **28**, 1–15 (1991)
13. Gupta, S.: Impact of sales promotion on when, what, and how to buy. J. Mark. Res. **25**, 342–355 (1988)
14. Cox, D.R.: Regression models and life-table. J. R. Stat. Soc. **34**, 187–220 (1988)
15. Seetharaman, P.B., Chintagunta, P.K.: The proportional hazard model for purchase timing: a comparison of alternative specifications. Mark. Sci. **23**, 234–242 (2003)
16. Seetharaman, P.B.: Additive risk model for purchase timing. J. Bus. Econ. Stat. **21**, 368–382 (2004)
17. Allenby, G.M., Leone, P.L., Jen, L.: A dynamic model of purchase timing with application to direct marketing. J. Am. Stat. Assoc. **94**, 365–367 (1999)
18. Ferreira, P.E.: A Bayesian analysis of a switching regression model; known number of regimes. J. Am. Stat. Assoc. **70**, 370–374 (1975)
19. Geweke, J., Terui, N.: Bayesian threshold autoregressive models for nonlinear time series. J. Time Ser. Anal. **14**, 441–454 (1993)
20. Chen, C.W.S., Lee, J.C.: Bayesian interference of threshold autoregressive model. J. Time Ser. Anal. **16**, 438–492 (1995)
21. Terui, N., Ban, M.: Modeling heterogeneous effective advertising stock using single-source data. Quant. Mark. Econ. **6**, 415–438 (2008)
22. Terui, N., Danaha, W.D.: Estimating heterogeneous price thresholds. Mark. Sci. **25**, 384–391 (2006)

23. Spiegelhalter, D.J., Best, N.G., Carlin, B.P., van der Linde, A.: Bayesian measures of model complexity and fit. J. R. Stat. Soc. Ser. B **64**, 583–639 (2002)
24. Newton, M.A., Raftery A.E.: Approximate Bayesian inference with the weighted likelihood bootstrap. J. R. Stat. Soc. Ser. B **56**, 3–48 (1994)

The Characteristics of Service Efficiency and Patient Flow in Heavy Load Outpatient Service System

Yuan Xu, Xiaopu Shang[✉], Hongmei Zhao, Runtong Zhang,
and Jun Wang

School of Economics and Management, Beijing Jiaotong University,
Beijing 100044, China
{17120627, sxp, rtzhang, 14113149}@bjtu.edu.cn,
zhaohongmei81@163.com

Abstract. In China's heavy load hospitals, the number of patients is far exceeding the hospital's service resources, and the hospital's outpatient system is usually much more complicated so that patients need to go through multiple stages to see doctors. In this study, based on the exploratory data analysis, we analyze the relationship among physician's service efficiency, the length of patient queue and the patient waiting time. The result indicates that the physicians' service efficiency has a positive correlation with the length of patient queue. The study also reveals that there is a certain correlation among the number of registered patients, the trend of change in the efficiency of doctor services, and the queue length of patients' wait in a day, and there is an effect of time lag among them, which could affect the efficiency of the treatment of the following stages by adjusting the treatment efficiency of the former stages. This work is aimed at the research of heavily loaded hospital outpatient systems. It is a result of exploratory data analysis on a large amount of real data and the relationship among the three variables mentioned above can assist the hospital in making decisions to some extent.

Keywords: Large hospital · Outpatient system · Service efficiency
Patient flow · Data driven

1 Introduction

Improvement on the hospital operation and management have been highly concerned by researchers, these researches include the uncertainty problems of outpatient visits [1], appointment issues [2], bed optimization [3], operation scheduling [4], and emergency department management [5], etc. Study on the patient flow and service efficiency is also an important topic, especially in the hospital which is a heavy load service system. There have been many studies on the patient flow. Ben-Tovim et al. [6] studied the flow of patients in the hospital by way of simulation, in which the key contribution is the rule discovery on the whole process of patient visiting. Yankovic et al. [7] proposed a patient queuing model considering the resources of beds and nurses, aimed to find the bottlenecks of the inpatient department, and the results of this

© Springer Nature Singapore Pte Ltd. 2018
J. Chen et al. (Eds.): KSS 2018, CCIS 949, pp. 17–30, 2018.
https://doi.org/10.1007/978-981-13-3149-7_2

study can be extended to emergency wards. Zacharias et al. discovered that the strategy of taking the appointment on the day of the visit can improve the efficiency of outpatient services based on the appointment outpatient system of US hospitals. This is different from the "walk-in" service mode [8]. Shi et al. [9] took a Singapore hospital as an example and proposed a "dual-time dimension" model to describe the residence time inpatients and simulated these two dimensions scenarios. Rohleder et al. [10] simulated and optimized the waiting time of outpatients from the microscopic level, which means the patient is concerned with individual patient other than the patient flow. This working result is suitable for relatively small numbers of patients, such as the private clinics, but cannot be used in outpatient system of large hospitals with heavy service load. In terms of hospital service systems, Qu et al. [11] proposed a two-section reservation model based on the integer programming approach. Since hospital patient flow is actually the dynamic changing which concerns with the arrival rate and service rate, many scholars have studied it from the perspective of queuing theory, and most of them assumed that the arrival of patients follows the Poisson distribution. Based on this, Baron et al. [12] proposed that adjusting the service resources to ensure that customers have a similar waiting time in different sections was able to bring customers a higher satisfaction degree. In fact, in large-scale hospitals, patient queuing involves many different sections, and it is hard to achieve a systemic balance without powerful simulation tools. Jiang [13] classified the outpatients and allowed them to perform multiple examinations at the same time to achieve parallel operations in some sections. For large hospitals with heavy loads, parallel examination consumes a lot of hospital service resources and it is difficult to estimate whether it can improve the treatment efficiency in large hospitals from the perspective of hospital congestion. There are also some empirical researches concluded that patients arriving in large hospitals do not comply with the Poisson distribution and there is no relatively stable state within a day [14]. Therefore, it is difficult to directly apply the traditional queuing theory into a heavy load "walk-in" outpatient system. Some studies suggest that the status of the queuing system and the patient's arrival rate are time-varying, based on which, some studies proposed a queuing model with the considerations that the time distribution of the arrival of the patients presents a time-related periodicity but assumed that the patient's service time is an average with an ideal distribution [15–17]. Based on the study result of Shi et al. [9], Dai [18] proposed a Mperi/Geo2timescale/N queue model for hospital inpatient flow from the perspective of queuing theory, which means the patient's arrival obeys a periodic Poisson distribution and service time is divided into two modules: "day" and "hour", and regard different wards in different departments as N parallel service desks.

In addition, some studies have found that doctor's service time and service efficiency in the hospital system may be related to the number of patients queued in the system and the degree of congestion [19, 20]. In this type of research, the general explanation for this phenomenon is that the pressure of the workers will be virtually increased when there are many patients with long queues or in service, which will have a positive effect of improving service efficiency, but when the service efficiency of medical staff may show a downward trend when the service intensity exceeds a certain level, which named an inverted "U-shape" or show a fluctuated "N-shape" [21, 22]. The above findings are all related to the hospitalization system and physicians' efficiency is mainly measured by the length of patient stay. In addition, the proportion of hospital beds is used to measure the work pressure of doctors in these studies. However, there is a stronger real-time nature of the physicians' pressure and service efficiency in the outpatient system, it belongs to the only dedicated work in a period, which are the different aspects from the inpatient systems. In this paper, we take the relationship between the patient flow and other aspects of the hospital outpatient procedure into consideration and aimed to optimize the service efficiency of the entire outpatient system.

The rest of this paper is organized as follows. Section 2 introduces the research question and data source. Section 3 analyzes the patient visiting procedure in outpatient system. Section 4 analyses the relationship among the different departments and puts forward some suggestions on the system optimization. Section 5 summarizes this work.

2 Problem Description

The efficiency of hospital service is closely related to the number of hospital staffs, and probably related to the number of patients. In a heavy load hospital, the service resources are very limited. Many scholars and hospitals are focusing how to make better use with the exiting medical resources and provide better service for patients. However, there are many sections in outpatient system, and the service condition and service efficiency of each section is closely related with the others. Therefore, to improve the efficiency of the whole outpatient system, it is necessary to outline the service features of each section.

This paper explores how the entire outpatient procedure changes after the patient entered the hospital from the perspective of the patient flow. This study regards hospital outpatient procedures formed by each section as a service network from the perspective of the system and each section is a sub-service system in the network. Based on the exploratory data analysis, this research analyzes the operating characteristics and laws of outpatients in large-scale hospitals through the collection and analysis of many patient flow data in the hospital information system to provide services for hospital operations management and decision-making.

The data of this study is from a regional central hospital of China, and the data has been processed to get rid of personal information. The items of the data include Patient ID (encrypted format), Doctor ID (encrypted format), Department ID, Patient registration time, the time of enqueue (patient), the time of entering consulting room (patient), the time of finish consulting (patient). In this study, we assume that (1) the outpatient system including the general outpatient department and emergency department. In this paper, we only focus the general outpatient. (2) Although some of the departments in outpatient system open on weekends, we only consider the running condition on working days. (3) The doctors can be divided into specialists and general physicians, and the general physicians visited by patients constitute the most part of visiting. In this paper, we only consider the service provided by general physicians.

3 Analysis of Outpatient Service Characteristics

Registration, payment, medical treatment and some medical examinations are included in Outpatient system, usually, patients' input, output and physicians' efficiency of the previous section are closely linked with the latter one, and the treatment process is shown in Fig. 1.

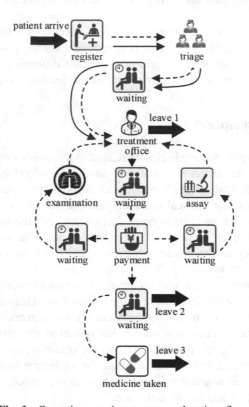

Fig. 1. Outpatient service process and patient flow.

As shown in Fig. 1, patients in outpatient system will experience "convergence" and "diversion" at different sections, and not all the patients have to go through all the sections, part of them may leave in the middle for some reason.

The treatment process consists of multiple sections. Therefore, the stay of patient in the hospital is the sum of the stay in all the sections. The treatment process in different departments is similar, mainly including registration, payment, treatment and medicine taking. There are some interactions among them, the common ones are "payment-examination" and "payment-medicine taken". As shown in Fig. 1, since these patients entered the hospital, the first diversion occurred after the patients completed the treatment. After that, some patients leave directly for some reason. The patients who enter the payment section are re-divided after it is completed. Part of them enter the "assay-treatment" section, part of them enter the "examination-treatment" section, the rest of the patients left the hospital directly.

Patients spend a great deal of time waiting for the admission to the next section in heavy load hospital outpatient. Therefore, what we should do is to try our best efforts to reduce the waste of these parts of time to achieve a maximum of physicians' service efficiency based on existing resources, so that the time patients waste in queueing for the next section can be reduced as little as possible.

The object of this study is a heavy load large hospital outpatient system. For example, in Fig. 2(a), there are five registration windows are in service status, at the same time, there are much more patients queueing in the windows, then we regard the scene as a heavy load system. On the other hand, if the number of the queueing patients is equal to or smaller than the windows in service, we see it as a non-busy system as shown in Fig. 2(b). Because the heavy load hospital outpatient system has a high probability of some patients arriving at the hospital at the same time, it difficult to tell when the patients arrive at the hospital through the timestamp data. Therefore, the patient arrival rate in heavy load outpatient systems do not obey Poisson distribution assumptions commonly used in current queuing system studies.

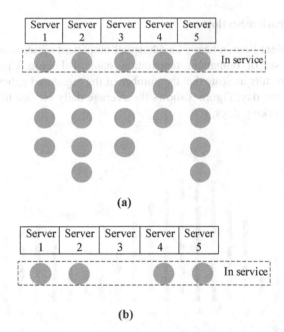

(a)

(b)

Fig. 2. (a) Heavy load service scenario with a large number of patients. (b) Service scenario with a small number of patients.

In heavy load outpatient systems, the number of service windows at each section is much smaller than the number of patients waiting in line, so the efficiency of each window is closely contact to the efficiency of the next section.

4   Analysis of the Relationship Between Outpatient System Sections

As mentioned above, outpatient system has many sections with inextricable links among them. Due to the limit space, this paper mainly describes the registration and treatment sections in detail. The registration is the first step for patients to enter the outpatient system. It's a necessary step that all the patients must do before they visit the doctor, so the registration is almost the busiest part of the outpatient system. Patients in the registration section do not involve the question of triage to different departments. Therefore, it can describe the whole hospital's patient flow to some extent. Between the registration and the treatment section, patients should go to the nurse station and are divided into different departments for treatment. During this process, patients should wait for being admitted to the doctor, which is the most time-wasting section. This study mainly analyzes these two sections and aims to discover the characteristics of the service efficiency and patient flow to reduce the time that patient stay in outpatient system and help heavy load hospitals to improve their operational management capabilities.

4.1   The Registration Section

To analyze the patient flow in heavy load hospital outpatient systems, this study makes a statistical analysis of outpatient registration data of Peking University Peoples Hospital, and the results indicate that the number of the registered patients is constantly fluctuating during the day. Figure 3 shows the average daily service times for different periods of 2016 working days.

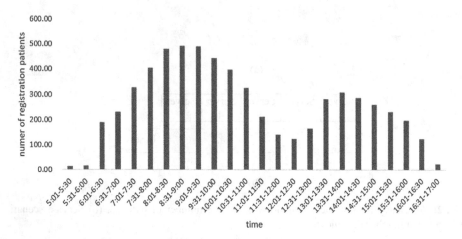

Fig. 3. Daily average number of patients' registration service in every 30 min

From Fig. 3 we can see that the busiest time of the registered window is 9:00 am and 2:00 pm, there are fewer registered people before and after this point, and does not show a relatively stable interval in the figure, which means the steady state does not exist.

The uncertainty of patients arrival in heavy load hospitals has made it difficult to estimate the arrival time of their patients. Therefore, in this study, a statistical analysis of the number pf people registered in different departments is used to observe the registration status of different departments at different times during the day, which is showed in Fig. 4. We can see that the number of patients reached a peak in the period from 8:00 to 9:00 am and 1:00 to 2:00 pm in almost all departments we chose in this study, which presents a "double-peak" structural curve.

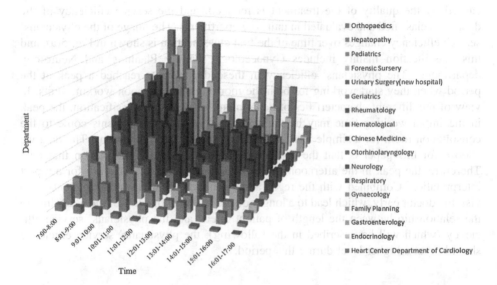

Fig. 4. Registration status at different departments in every one hour

In the hospital, each department has its own characteristics, so the management mode of different departments will also be different. Considering from the subjective level, the "double-peak" curve may be related to the fact that the number of patients in heavy load hospitals is usually bigger, and patients usually choose to register at the first time when the hospital staffs start to work. Therefore, there are usually more patients at the beginning of a work unit. The "double-peak" curve of the registration is an important feature and have a great influence to the efficiency of the next section.

4.2   The Triage and Treatment Sections

After the triage, the patients enter different departments. This study selects some departments in combination with the characteristics of the Peking University Peoples Hospital and classifies them by calculating the service efficiency of each department doctor.

We calculate the physicians' efficiency who is working in different departments based on the timestamp data of the time of finishing consulting in the hospital system, and then divide the departments into the following categories.

As we know, in addition to the length of patient queue, the physicians' service efficiency is closely related to many factors, such as the quality of service, the type of department, the patient's condition, and even the quality of the patient. This study discusses the service efficiency of each stage of the outpatient clinic based on the timestamp data in the hospital information system. Therefore, we define the service efficiency as the flow rate of outpatients at different stages. Through the measurement of the overall flow of patients in different departments, the uncertainty of the efficiency caused by the quality of the treatment is reduced, and the service efficiency of the doctors is classified and evaluated in units of departments. The image of the physicians' service efficiency changes over time of the first classification is shown in Fig. 5(a), and this classification mainly includes Gynaecology, Family Planning and Neurology departments. The physicians' efficiency in these departments reached a peak at the period when they start working for both the morning and afternoon working units. In view of the different characteristics of department visits in this classification, the peak in the initial working time may be the result that quite a few patients come to the consultation room for a simple visit and then go for some examinations. The data we choose for this paper is that the results of examination can come out on that day. Therefore, the peak in the afternoon can be explained as patients' disturb for a report interpretation. Combined with the registration situation, there are many patients who visit the doctor early, which lead to a longer queue length at that time. And according to the relationship between the length of patient queue and the physicians' service efficiency (which will be described in the following), the physicians' service efficiency should be at a higher level during this period.

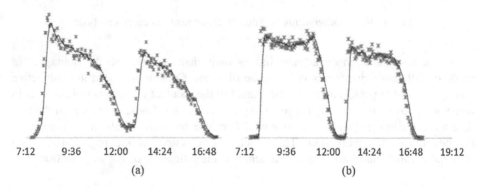

7:12   9:36   12:00   14:24   16:48   7:12   9:36   12:00   14:24   16:48   19:12

(a)   (b)

Fig. 5. Daily average number of patients' registration service in every 30 min

The image of the physicians' service efficiency changes over time of the second classification is shown in Fig. 5(b), and this classification mainly includes Paediatrics, Otorhinolaryngology, Rheumatology and Immunology, endocrinology and Heart Center Cardiology departments. The characteristics of these departments are that physicians' service efficiency has a relatively stable intervals both in the morning and afternoon of the day, and there is a certain degree of decline in their respective ranges. We dare to suppose reasons for the difference between the Fig. 5(a) and (b) is that the departments in two categories have different examination items and quantities, patients in (a) have more X light examinations while patients in (b) have more Ultrasonic examinations, and the average time required for X-ray examination is longer than that of Ultrasonic. Therefore, the time for the patients to return to the consulting room in Fig. 5(b) is shorter than that in Fig. 5(a), so we can see the relatively stable tendency in Fig. 5(b).

Other departments, such as orthopedics, general surgery, and gastroenterology departments have a relatively stable range of physicians' services in the morning, and an extraordinary upward trend of physician's efficiency in gastroenterology. Due to space limitations, these departments are no longer described in this study.

Figure 6 reflects the length of patient queue changes over time from 7:30 am to 5:30 pm. From Fig. 6 we can see that the queue length reaches a peak around 9:30 am and 2:30 pm in most departments except the light orange line which represents the Heart Center Cardiology department has a relatively smooth interval in the morning.

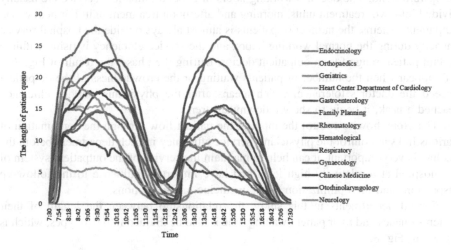

Fig. 6. The length of patient queue changes over time in different departments

By calculating the relationship between the length of patient queue and physicians' service efficiency, we obtain the image shown in Fig. 7. In the heavy load hospital outpatient system, the physicians' service efficiency increases with the length of patient queue, which explains the invisible correlation among the number of patient registration, the length of patient queue and the physicians' service efficiency from the perspective of data. Since these three are in different sections, there are some effects of time lag, manifested in the peak number of registration patients, the peak the length of patient queue and the peak of physicians' service efficiency has a certain time dislocation.

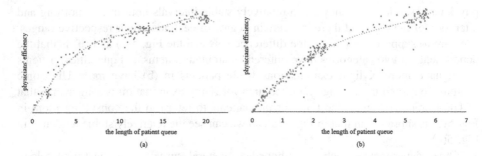

Fig. 7. Physicians' efficiency changes over the length of patient queue wait

For different departments, the relationship between the length of patient queue and physicians' service efficiency can be roughly divided into two types and both of which are positively related. The difference between them is that when the length of patient queue within the available data reach a certain length, the physicians' service efficiency in the first type of departments shows a declining trend, while the second type still has an upward trend. The doctor's working hours in general outpatient service are usually divided into two treatment units, morning and afternoon treatment unit. For heavy load outpatient systems, the number of patients is almost always outside the hospital service capacity during the normal working hours, and the service efficiency is rising within a certain patient range for an outpatient doctor. During the phase, the result of Fig. 7(a) will appear when the number of patients waiting or the crowdedness of the outpatient exceed the doctor's tolerance, which means after the physicians' service efficiency reached a peak, it begin to show a downward trend.

Therefore, how to control the input of the patient flow through the combination of various links to control the physicians' service efficiency in each department before the decline is very important, it can help to maintain the service of the outpatient system of the hospital at a relatively high level and have important influence to the follow-up inspections and the optimization of the medicine taking sections.

For all departments in this study, the relationship between the length of their patients queues and their patient waiting time is also be divided into two types, which is shown in Fig. 8.

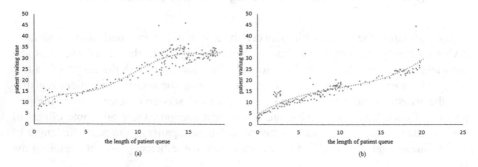

Fig. 8. Wait time for treatment changes over the length of patient queue wait

From Fig. 8, we can see that the relationship between the length of patient queue and their waiting time is positively related. Figure 8(a) represents the departments in which the patients' wait time stays relatively stable when patients queue reaches a certain length and afterwards there is still a rising trend, like Otolaryngology, Traditional Chinese Medicine departments. Figure 8(b) represents the departments in which the patients' wait time is almost linear with the length of the queue, like Liver Diseases Branch, Geriatrics, Neurology and Gastroenterology, etc.

For the first type of department, there is a relatively stable interval regardless of the grow length of the queue, and based on the Fig. 7 that the relationship of the physician's efficiency and the length of patient queue is positively related, we can improve the hospital's service efficiency by increasing the patient input flow to get a higher physicians' service efficiency and a lower patient waiting time for treatment. For example, try to keep the patient waiting time during the visit stage in a relatively stable state with an increase in the patient cohort by increasing the speed of patient flow during the triage stage. For the second type of department, we need to analysis the characteristics of the specific departments to find some relationships. Because the different departments have their own characteristics, on the basis of the major categories of the department, it must according to their respective characteristics for further analyze. Due to the space limitations, this study does not describe in detail, and it will be a complex but meaningful work.

4.3   Analysis of the Relationship Among Different Sections

Patients need to go through several sections in hospital outpatient services. It is vital for hospital operators and managers to make decisions about how to place the limited resources to the key section scientifically and reasonably so that they can improve the service efficiency of the whole hospital. The patient flow put these sections together, and where the input patient flow have a great influence on the output patient flow of the next section. In the following, we abstract out some functions to describe the relationship among time, physicians' service efficiency, the length of patient queue and the patient waiting time involved in several sections more clearly.

The function of patient registration number changes over time is denoted by $R(t)$; the function of the length of patient queue changes over time is denoted by $Q(t)$; the function of physicians' service efficiency changes over time is denoted by $E(t)$; the function of physicians' service efficiency changes over the length of patient queue is denoted by $E(q)$ and the function of the patient waiting time changes over the length of patient queue is denoted by $W(q)$, where t represents time in all the above functions.

We have mentioned that there is a time lag among the peak time of registration, service efficiency, the length of queue and waiting time and we definition it as Δt. If we set the patient registration as the initial time t_0, then we can obtain

$$R(t) = R(t_0), \tag{1}$$

$$Q(t) = Q(t_0 + \Delta t_1), \tag{2}$$

$$E(t) = E(t_0 + \Delta t_1 + \Delta t_2), \tag{3}$$

where Δt_1 represents the time lag between registration time and queue time, Δt_2 represents the time lag between queue time and the service time.

From the perspective of a single patient, they can predict their own treatment situation to the entire hospital, and from the perspective of the hospital, the hospital's service can be reflected by the size of Δt and can be improved by reducing Δt of the former section.

5 Conclusions

This study use the data-driven method as the fundamental thought to calculate the service efficiency of different departments of the hospital outpatient service system accurately and combine the patients flow from the registration section to evaluate the outpatient service efficiency, and then classify the departments based on the efficiency characteristics to make a pertinent discussion and provide some relevant recommendations.

In this study, based on exploratory data analysis, the different treatment sections of different departments are divided into several categories, and the relationship among physicians' service efficiency, the length of patient queue and the patient waiting time is discovered. The physicians' service efficiency increased with the length of patient queue grows and the patient waiting time and the length of patient queue also showed a positive correlation. The study found that for some departments, the patient's queue length and physicians' efficiency, patient's waiting time are not a strictly positively related. When the patient's queue is greater than a certain length, the physicians' workload exceed the bearing capacity and lead to a decline in the efficiency of the work, and those departments where do not exist these problems may be due to the conflict between the existing medical resources and the number of patients in the hospital is not very prominent. However, it does not exclude that this situation will appear when the patient flow increases some time in the further. When the patient length is within a certain range, the impact on patient's waiting time is not obvious in certain departments. This study breaks down the traditional thought that can only increase the efficiency of hospital services by increasing medical resources, and analyses different departments with the method tracking the patient flow to improve the physicians' service efficiency and provides an important reference for the management theories of the informatized hospitals.

This study only discusses the service efficiency of each stage of the outpatient clinic based on the time stamp data in the hospital information system, and only considers the influence of the length of patient queue on physicians' efficiency among different departments. In the following studies, we will consider more sections of the outpatient system, including the examination, medicine taken and payment sections, and analyze the patient flow on physicians' service efficiency from the perspective of the entire outpatient system, and based on which simulate a hospital's operating environment to help the hospital conduct their operational management better.

Acknowledgements. This work was partially supported by National Science Foundation of China (Grant number 61702023, 71532002), Humanities and Social Science Foundation of Ministry of Education of China (Grant number 17YJC870015), the Fundamental Research Funds for the Central Universities of China (Grant number 2018JBM304), the Fundamental Research Funds for the Central Universities (Grant number 2017YJS075).

References

1. Harris, S., May, J., Vargas, L.: Predictive analytics model for healthcare planning and scheduling. Eur. J. Oper. Res. **253**(1), 121–131 (2016)
2. Riise, A., Mannino, C., Burke, E.: Modelling and solving generalised operational surgery scheduling problems. Comput. Oper. Res. **66**, 1–11 (2016)
3. Harper, P., Shahani, A.: Modelling for the planning and management of bed capacities in hospitals. J. Oper. Res. Soc. **53**(1), 11–18 (2002)
4. Li, F., Gupta, D., Potthoff, S.: Improving operating room schedules. Health Care Manag. Sci. **19**(3), 261–278 (2016)
5. Abualenain, J., Frohna, W., Shesser, R., Ding, R., Smith, M., Pines, J.: Emergency department physician-level and hospital-level variation in admission rates. Ann. Emerg. Med. **61**(6), 638–643 (2013)
6. Ben-Tovim, D., Filar, J., Hakendorf, P., Qin, S., Thompson, C., Ward, D.: Hospital event simulation model: arrivals to discharge–design, development and application. Simul. Model. Pract. Theory **68**, 80–94 (2016)
7. Yankovic, N., Green, L.: Identifying good nursing levels: a queuing approach. Informs **59**(4), 942–955 (2011)
8. Zacharias, C., Armony, M.: Joint panel sizing and appointment scheduling in outpatient care. Manage. Sci. **63**(11), 3978–3997 (2016)
9. Shi, P., Chou, M., Dai, J., Ding, D., Sim, J.: Models and insights for hospital inpatient operations: time-dependent ED boarding time. Manage. Sci. **62**(1), 1–28 (2015)
10. Rohleder, T., Lewkonia, P., Bischak, D., Duffy, P., Hendijani, R.: Using simulation modeling to improve patient flow at an outpatient orthopedic clinic. Health Care Manag. Sci. **14**(2), 135–145 (2011)
11. Qu, X., Peng, Y., Kong, N., Shi, J.: A two-phase approach to scheduling multi-category outpatient appointments – a case study of a women's clinic. Health Care Manag. Sci. **16**(3), 197–216 (2013)
12. Baron, O., Berman, O., Krass, D., Wang, J.: Using strategic idleness to improve customer service experience in service networks. Oper. Res. **62**(1), 123–140 (2014)
13. Jiang, L., Giachetti, R.: A queueing network model to analyze the impact of parallelization of care on patient cycle time. Health Care Manag. Sci. **11**(3), 248–261 (2008)
14. Brahimi, M., Worthington, D.: Queueing models for out-patient appointment systems—a case study. J. Oper. Res. Soc. **42**(9), 733–746 (1991)
15. Green, L., Kolesar, P., Whitt, W.: Coping with time-varying demand when setting staffing requirements for a service system. Prod. Oper. Manag. **16**(1), 13–39 (2007)
16. Mardiah, F., Basri, M.: The Analysis of appointment system to reduce outpatient waiting time at Indonesia's public hospital. Hum. Resour. Manag. Res. **3**(1), 27–33 (2013)
17. Allon, G., Deo, S., Lin, W.: The impact of size and occupancy of hospital on the extent of ambulance diversion: theory and evidence. Oper. Res. **61**(3), 544–562 (2013)
18. Dai, J., Shi, P.: A two-time-scale approach to time-varying queues in hospital inpatient flow management. Oper. Res. **65**(2), 514–536 (2017)

19. Kc, D., Terwiesch, C.: Impact of workload on service time and patient safety: an econometric analysis of hospital operations. Manage. Sci. **55**(9), 1486–1498 (2009)
20. Kc, D., Terwiesch, C.: An econometric analysis of patient flows in the cardiac ICU. Manuf. Serv. Oper. Manag. **14**(1), 50–65 (2011)
21. Findlay, M., Grant, H.: An application of discrete-event simulation to an outpatient healthcare clinic with batch arrivals. In: Proceedings of the 2011 Winter IEEE Simulation Conference (WSC), pp. 1166–1177 (2011)
22. Jaeker, J., Tucker, A.: Past the point of speeding up: the negative effects of workload saturation on efficiency and patient severity. Manage. Sci. **63**(4), 1042–1062 (2017)

How Do You Reduce Waiting Time?
Analytical Model Expansion with Unstructured Data

Keiichi Ueda[✉] and Setsuya Kurahashi

University of Tsukuba, 3-9-21 Otsuka, Bunkyo, Tokyo, Japan
s1645001@u.tsukuba.ac.jp
http://www.gsbs.tsukuba.ac.jp/en/

Abstract. Numerous studies have investigated the factors that influence decision-making. Questionnaires have played a critical role in verifying the statistical significance of the conceptual model in these studies. In addition to observable facts, it is extremely important that records of memories obtained from surveys accurately describe the judgment process of individuals. However, it is also a fact that subjective memory is not always objectively accurate owing to indications that marketing research methods for understanding consumers alter their memory. In general, data on individual decision-making in a service operation does not remain in the record. In this study, we examine whether unstructured data from games can effectively contribute to service improvements using a consumer's decision-making model of service selection. The agent-based decision-making model is expanded to a gaming framework. We determine that the clues, obtained through cooperative games based on this model, effectively contribute to service improvements. Players discuss their experiences of the game during a debriefing. The extracted strategy that uses the unstructured data, such as the awareness and discussion outcomes of players, is examined through computer simulation.

Keywords: Self-service technology · Agent-based modeling · Airport
Game · Simulation · Service · Knowledge

1 Introduction

1.1 Background

An aging society with fewer births and the continuous growth of the service economy have resulted in imminent and ongoing issues. Developed countries are currently challenged with the problem of securing services for their workforce. We expect that self-service technologies (SST) will play an important role in fulfilling future customer service requirements. However, successful implementation of SST is not achievable unless SST is recognized and accepted by customers and employees. Being aware of people's perceptions of SST can be helpful but difficult to understand. Zaltman claimed that 95% of the decision-making process occurs below the conscious level. He indicated that knowledge of consumer

© Springer Nature Singapore Pte Ltd. 2018
J. Chen et al. (Eds.): KSS 2018, CCIS 949, pp. 31–44, 2018.
https://doi.org/10.1007/978-981-13-3149-7_3

decision-making that is determined through telephone interviews, focus groups, and questionnaires do not accurately explain the manner in which people think [25].

This study focuses on a self-service kiosk at an airport and discusses the ways in which the service provider utilizes service resources, including SST. The core product of an airline includes simultaneous and inseparable services. Our research focuses on passenger handling to consider the ways in which service resources can be optimized through the utilization of our knowledge and unstructured data. Service operation in a departure lobby must manage service agents, service facilities, passengers, and a self-service kiosk. As the check-in process is a critical starting point of the travel experience, service staff and passengers must work together to achieve a common goal in each process of the air travel experience.

It is essential for airlines and passengers to reduce the amount of waiting time. However, as airlines cannot predict the floor through which passengers will arrive, it is cumbersome to manage service resources for handling ever-changing situations. Some assistance is required for local managers to explore better service operations and evaluate such operations quantitatively and qualitatively. However, an examining methodology of the best practices with limited operational risks is yet to be developed.

1.2 Objective of This Study

In this study, we determine that a gaming framework contributes to the development of handling strategies in ever-changing situations involving multi-participants and several variables. The cooperative game is examined to determine a way of optimizing current resources of an airline. We discuss the method of extracting key ideas that effectively control the service level. The gaming framework incorporates the mechanism of coordination and cooperation based on the platform where passenger agents autonomously make decisions. The objective of this study is to utilize the knowledge and implications of handling complicated ever-changing situations. To achieve this, two steps are considered for uncovering the fundamentals of a smoother passenger handling operation: constructing the gaming framework and performing computer simulation under several scenarios.

2   Related Work

We briefly review four major areas to gain an understanding of situations involving service operations. Innovation diffusion and services marketing are examined to determine the basic ideas involving decision-making by individuals. An agent-based SST adoption model is reviewed because of its applications in reproducing the dynamics of service choosing. In addition, we use this model as a platform for gaming. We also briefly review the gaming literature, which describes the ways of handling complicated decision-making and the manner in which games should be designed by considering such complicated decision-making.

2.1   Innovation and SST Studies in Service Marketing

SST is described as a technological interface that enables customers to provide a service without the involvement of service employees [15]. It is widely recognized that SSTs can increase productivity and efficiency, while simultaneously contributing to the reduction of labor costs [9, 13, 14]. Rogers defined innovation as the introduction of something new: a new idea, method, or device [18]. This is accomplished using more effective products, processes, services, technologies, or business models that are readily available in markets, governments, and society. Greater relative advantages and the efforts of change agents are known to accelerate the diffusion of innovation [18]. Because using SST is an individual decision, studies pertaining to the adoption and diffusion of innovation provide many implications. Berry et al. examined and discussed convenience from two main perspectives: wait time and its management, and what consumers find convenient [1]. Davis developed a "technology acceptance model" (TAM) that is specifically intended for explaining computer usage behavior [6]. The methodology developed emphasizes the necessity of evaluating new proposed systems prior to their implementation. He also empirically examined the ability of TAM to predict and explain user acceptance and rejection of computer-based technology [7]. Meuter et al. elucidated that when SST is better than alternatives and when consumers greatly appreciate savings on time, service convenience through SST results in higher consumer satisfaction [15]. Bitner et al. focused on the benefits of thoughtfully managed and effectively implemented technology applications. They emphasized the ways in which service encounters can be improved through the effective use of technology [2]. Customer readiness for SST is examined as a dependable variable [3, 13], and technology anxiety is also explicated and identified as a superior, more consistent predictor of SST usage than demographic variables [14].

Dabholkar and Bagozzi extended the attitudinal model of technology-based self-service and proposed that moderating variables such as consumer traits and situational factors affect attitudes toward SST and the intention of using SST [5]. Gelderman et al. explicated that technology readiness has no impact on customer decision-making and specified that perceived crowdedness exhibits a strong and significant impact on the decision of customers to use SSTs [9].

2.2   Agent-Based Modeling: SST Adoption Model

Agent-based models (ABMs) represent the phenomena of complex social systems [12]. ABMs enable each heterogeneous agent to behave autonomously and allow interactions to emerge in the experimental space, which is approximated to the real world. ABMs are useful for problems involving emergence because they are composed of multiple levels and interactions. ABMs are different from traditional models because of their complexity, which includes representing how individuals and the environmental variables that affect them vary over space and other dimensions such as time [16]. Stylized facts are one of the many ways of ABM validation. They are empirical regularities in the search for theoretical and causal explanations such as statistical features [24].

The emulated airport lobby by ABM is demonstrated by Ueda and Kurahashi (Fig. 1). They proposed the decision-making mechanism of SST use [21] in ABM and expanded the mechanism by incorporating individual traits [22]. The experiment supports the representation of actual results under several different conditions. Cooperation between passenger and service staff and coordination among service resources are examined and explicated. The model is validated at different hierarchical levels [22,23].

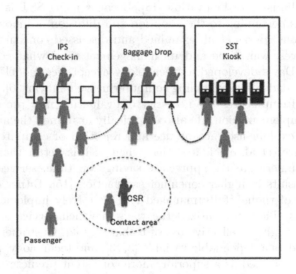

Fig. 1. Experimental space of the airport departure lobby

2.3   Gaming and Simulation

Representing the complicated phenomena involved in decision-making by people using computer simulation can be challenging. Gaming is one of the ways of handling continuously changing situations and according judgment of player participation in complicated events [20]. Richard claimed that because role-playing allows multiple stakeholders to examine complex systems, it can help in conveying complexities through communication [8]. Gaming is considered a useful tool for learning to deal with complicated real-world issues. Many educational settings, such as corporate and military training environments, embrace games and simulations [17]. Salen et al. claims that all games include a core mechanic for designing the player's expected experience. The core mechanic defines an action or a set of actions that players will repeat several times as they move through the designed system of the game [19]. Gaming provides opportunities to learn about responses in the real world. Greenblat proposed the design process of a game that is similar to the construction of an ABM [10].

3 Constructing the "Departure Lobby Management Game"

3.1 The Core Mechanic of the Game

We expand the agent-based SST adoption model into a business game and explain how it is employed in gaming. This game has an ambivalent structure in which an increase in the waiting time of one area leads to a deterioration of the service level in another area. There are two expected findings that are described in the core mechanic (Fig. 2): the structure of the game for determining key points that make maximum use of service resources; and the ways of handling real responses and endings that are observed in the workplace through subjective experience.

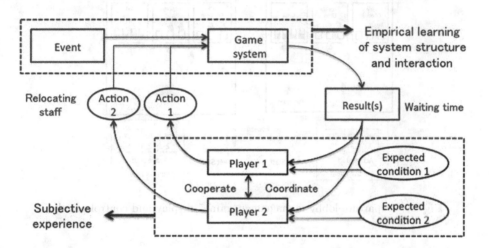

Fig. 2. The core mechanic of the "departure lobby management game"

3.2 Gaming Outline

Game Players and Variables. There are two separately managed departure lobbies (Fig. 3). Each lobby includes facilities, equipment, and a responsible manager. The manager of lobby-A oversees nine staff members and the manager of lobby-B manages six staff members for operating service facilities. The managers locate their staff at check-in positions (CC), baggage dropping position (BD), and in the lobby (CSR) to guide and advise passengers on how to use a self-service kiosk. Each manager can utilize staff who are resting. However, the manager of lobby-A can retrieve their staff to activate the service facility of lobby-A at any time. The arriving passenger volume and timing is completely different in each lobby. 30% of lobby-A and 70% of lobby-B passengers must have their baggage checked.

Trade-Off in Gaming. Heterogeneous passenger agents autonomously select service options. They recognize their surrounding environment and decide their walking direction. CSR's interaction reduces a passenger's hesitation, which increases the passenger's possibility of heading to the SST. Each passenger counts the waiting time as soon as they join the line in front of the service facility. When the waiting time exceeds a certain threshold, the color of the passenger turns red and the internal status changes to "angry." After queuing for a check-in, passengers in red are redirected to the other lobby. This replicates the situation where passengers in a long queue will miss the flight and require re-booking. "Angry" passengers from the other area take more time to process re-booking and check-in. Consequently, this increase the waiting time for other passengers.

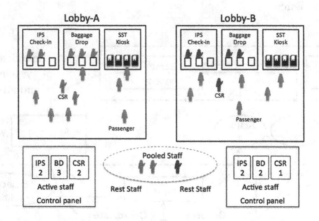

Fig. 3. "Departure lobby management game" display and control panel

4 Experimental Results

4.1 Gaming Results

The cooperative gaming experiment was performed after participants were familiarized with controlling parameters and the gaming concept. The actual passenger arrival curve, which is completely different, is used in the departure lobby. Players must observe and predict passenger congestion and then determine the actions that must be undertaken.

The games were conducted thrice to determine the ways in which players develop their strategy and teamwork. Figure 4 describes the total number of passengers in each lobby and the number of "angry" passengers along the time series (in ticks). Figure 6 illustrates that the change in the number of BD correlates with the appearance of "angry" passengers. The number of "angry" passengers increases at 1396 as shown in Fig. 4(a) Baggage Drop of lobby-B (BDb) decreases from 3 to 2 at 1174 (ticks) to 1383 (ticks) in game-1 as described in Fig. 6(a). The number of "angry" passengers rises to 1381 (Fig. 4(c)) BDb decreases to 2 at

1139 and 1308 in game-3 (Fig. 6(c)). However, in game-2, Fig. 4(b) demonstrates that the number of "angry" passengers increases to 1465 and then quickly stops increasing. This is most likely because BDb quickly turns 3 (at 1250) after BDb decreases to 2 positions at 1211 as indicated in Fig. 6(b) (Table 1).

Table 1. Gaming experiment: the number of "angry" passengers

Experiment	"Angry"	BDb = 2 Duration (ticks)	"Angry" span (ticks)
Game-1	25	210	1463
Game-2	10	40	962
Game-3	20	170	1427

Likewise in game-2 and game-3, Fig. 6(b) and (c) illustrate that three lobby service agents were located in lobby-B (CSRb) at 1083 (ticks) and 642 (ticks) before the first "angry" passenger appears at the timing of 1126 (ticks) and 682 (ticks) in Fig. 4(b) and (c). Because the game designs CSR to reduce hesitation toward SST, more passengers were encouraged to use SST in lobby-B. This consequently increased the number of passengers queuing at BD because a specific number of passengers operating SST needed to visit BD to check their baggage.

Findings of the Gaming Experiment. There are two findings from the review. Focusing more attention on the quantity of BD would yield better results. Players can determine the opening of the third BD by predicting passenger congestion more carefully. It is critical to note that SST users with baggage revisit BD before leaving the departure lobby. This consequently increases their waiting time because they need to join the queue at BD. As passengers in lobby-B have a higher baggage holding rate, more SST users with baggage result in an increase in waiting time, which leads to more "angry" passengers in total. Fewer CCRs relate to fewer SST users in lobby-B, which contributes to the reduction of "angry" passengers. This gaming result also implies that CSR plays an important role as a "change agent."

4.2   Computer Simulation Results

The computer simulations were performed using the same platform as for gaming. We prepared 21 scenarios to clarify whether the findings of gaming experiments are valid (Fig. 7). CC22BD33CSR21 denotes 2 CCs and 3 BDs as active in both lobby-A and lobby-B, and there are 2 CSRs in lobby-A and 1 CSR in lobby-B. The result demonstrates that the third BD is effective in reducing the number of "angry" passengers.

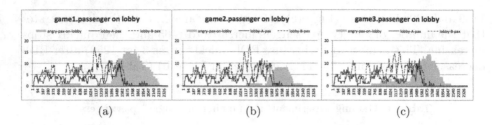

Fig. 4. Passenger appearance

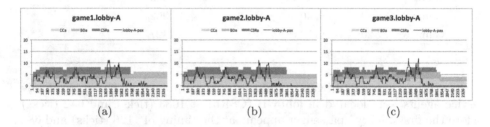

Fig. 5. Lobby-A operation process

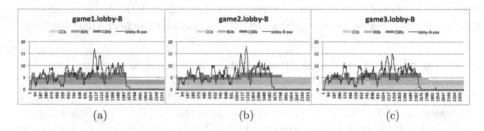

Fig. 6. Lobby-B operation process

In addition, assigning more CSRs and the number of "angry" passenger has a positive correlation under 3 CCs (Fig. 8). Assuming that 3 active CCs provide good capability for handling incoming passenger-agents, only a small number of passenger-agents consistently reach BD via CC. If more CSRs succeed in sending passengers to SST, more passengers will use BD for baggage check because 70% of the passengers have baggage. This results in increased congestion in BD waiting lines, which leads to increased "angry" status among passengers.

It is also observed that there is a nonlinear relationship between the number of CSR and "angry" passengers under 2 CCs in Fig. 9. There are two paths that passengers can take to reach BD: via CC and via SST. Passengers in CC waiting lines utilize BD when there are fewer passengers waiting in front of BD than those waiting in front of CC. The passengers can complete check-in at one time (obtain a boarding pass and check their baggage) with BD. It takes longer time than simple baggage check-in.

The computer simulation result indicates that passenger-agents overflow to BD from CC is consistent because of limited capabilities if the number of CC remains two. This overflow results in long queues and consequently more "angry" passengers. Guiding passengers to SST by assigning one CSR helps reduce passenger overflow to BD and reduces the total processing time in waiting lines of BD even if the number of passenger-agents via SST increases. If the reduced number of overflown passengers via CC go to BD after using SST within the same timeframe, the total processing time in BD lines is reduced. However, the simulation results (Fig. 9) indicate that assigning more than one CSR in respective lobbies increases the processing time of BD line as compared to the assignment of a single CSR.

These results support what players learned and their findings are critical for gaming experiments instead of performing considerable scenario analysis through computer simulation. Therefore, gaming can productively provide clues concerning complicated context such as responding to a situational change as time passes. The data obtained from gaming and computer simulations for evaluating ideas provides additional findings that lead to new concepts and extends the views. Newly obtained ideas may be incorporated to develop analytical model in conceptual and/or implemented model (Fig. 11).

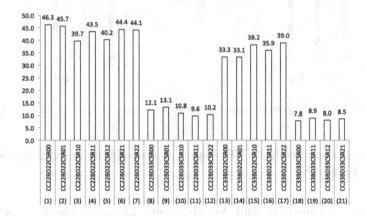

Fig. 7. Computer simulation results

4.3 Discussion and Summary of Gaming

Key Findings of Operational Strategy. It is critical to observe arriving passengers to predict the amount of waiting time so that players can determine whether to open the third BD. This gaming result also implies that CSR plays an important role in queue management. The gaming and computer simulation results demonstrate that more active BD position directly contributes to the reduction in the number of "angry" passengers. The computer simulation results indicate that the effect of CSR varies according to the handling capacity of CC. The total waiting time of each passenger and waiting time as a whole could have

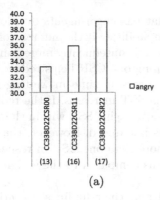

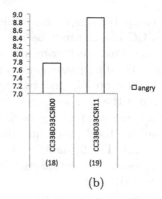

(a)   (b)

Fig. 8. Simulation results of active 3 CCs

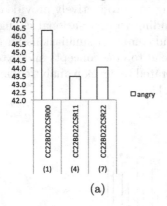

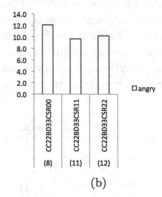

(a)   (b)

Fig. 9. Simulation results of active 2 CCs

been reduced dramatically if CSRs could intentionally invite only non-baggage passengers to use SST.

The Chain of Findings and Verification. These are the key findings of gaming that were neither focused on nor observed in the database. In other words, they are discovered from unstructured data. They are outcomes of dialog and discussion among game players. The findings enhanced analysis-discussion to determine a way of reducing the waiting time in the lobbies. The discussion helped players make in-depth analysis of the system, including the rise in the number of "angry" passengers.

The computer simulation with multiple scenarios then assumed a role in the validation of ideas. For verification, we incorporated newly obtained data elements in the analysis, which are structurally collected from the computer simulation.

Importance of Communication. The previously mentioned findings include the outcomes of communication during the gaming operation and a discussion on the debriefing of the game. The interactive questions and answers among the participants accelerate the process of understanding the system structure and interactions. The discussion also unveils the background of the decision-making process of participants. A proper debriefing can maximize learning outcomes [4]. These opportunities contribute to learning the "subjective experience" of other stakeholders.

5 Conclusion

Through this research, we examined the utilization of knowledge and awareness and the implications of the ways of handling complicated ever-changing situations with multiple participants.

5.1 Summary of This Work

The Analysis Model Development. Owing to the nature of the service, it is essential to discuss and verify the service quality level by co-working with the service recipient. In reality, it is difficult to quantify the service quality. In addition, the opportunity for learning this issue through actual operations is practically limited. This challenge is borne out of the fact that this issue must handle an ever-changing situation.

We reviewed the related interdisciplinary research on SST studies, which verify the concept model with a statistical method using quantitative data and visualized facts [5, 9]. SST research evolves into a dynamic model by incorporating passenger behavioral rules obtained from service experts [21] and influencing factors newly discovered from time series data [22]. The former was extracted from non-structured data and the latter was extracted from structured data.

Gaming Framework Stimulates Discussion. Based on the agent-based SST adoption model, which reproduces the real-world context, we developed the gaming framework to discuss the key factors in improving service by utilizing service resources including SST. By practicing the game, players explore and exploit different strategies and gain an in-depth understanding about what to do by participating in discussions.

The following two findings are extracted. First, it is possible to handle complicated issues related to human decision-making through games with limited amounts of unstructured data. The gaming participants can develop a strategy. Second, a strategy can be formulated that contributes to improved services without trying innumerable combinations according to time series or various situations. The newly selected elements are collected for verification. We verify the effectiveness of the strategy through computer simulations using the same platform.

Knowledge Expansion by Incorporating Unstructured Data. We claim
that communication through gaming experiments have resulted in the following
findings. The participants obtain a deeper understanding of practical issues and
how to deal with such issues through discussions. The debriefing is useful and
effective for formulating future strategies to enhance services. It helps players
uncover the system structure and interaction with the game. It promotes mutual
understanding and allows a consensus to be formed on a new handling policy.

In this study, we reiterated new findings and expanded the analytical model
by incorporating unstructured data (Fig. 10).

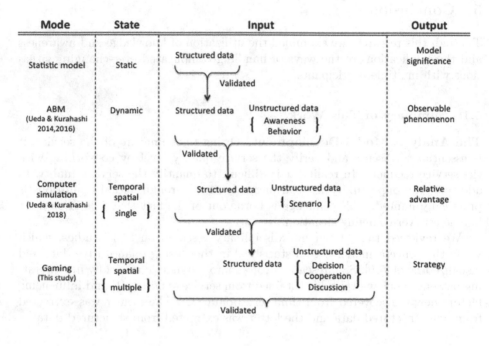

Fig. 10. Knowledge expansion with unstructured data

5.2 For Future Study

Limitation. In this study, we addressed the issues in the way of "Time and
Space Specified" [11]. We do not aim to reveal how people think. The satisfaction
level of the service recipient, which is the ultimate and essential objective of the
quality of service provision, is not covered under this research. The parameter
values of the platform model used in this experiment are exaggerated instead of
mapping the real world to emphasize the core mechanic of the game.

For Future Study. To provide more realistic learning opportunities, we would
require additional progress in incorporating factors and scales that better explain
the status of customer satisfaction. It would be better to modify the model in

terms of the times that passenger agents get "angry" and the addition of alternative CSR guiding policies. As the next step, we expect to accumulate more quantitative and qualitative data. The analytical framework can be developed by incorporating subjective experiences and obtaining data through more gaming experiments. By reflecting the outcome of experiments, we can expand the existing analytical model. This course of action (Fig. 11) will help explore empirical solutions in a complex world.

A   The Analytical Development Life Cycle

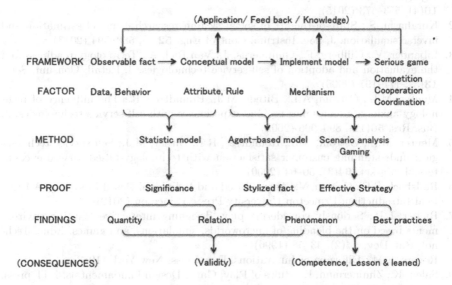

Fig. 11. The analytical development life cycle

References

1. Berry, L.L., Seiders, K., Grewal, D.: Understanding service convenience. J. Mark. **66**(3), 1–17 (2002)
2. Bitner, M.J., Brown, S.W., Meuter, M.L.: Technology infusion in service encounters. J. Acad. Mark. Sci. **28**(1), 138–149 (2000)
3. Bitner, M.J., Ostrom, A.L., Meuter, M.L.: Implementing successful self-service technologies. Acad. Manag. Exec. **16**(4), 96–108 (2002)
4. Crookall, D.: Engaging (in) gameplay and (in) debriefing. Simul. Gaming **45**(4–5), 416–427 (2014). https://doi.org/10.1177/1046878114559879
5. Dabholkar, P.A., Bagozzi, R.P.: An attitudinal model of technology-based self-service: moderating effects of consumer traits and situational factors. J. Acad. Mark. Sci. **30**(3), 184–201 (2002)

6. Davis, F.D.: A technology acceptance model for empirically testing new end-user information systems: Theory and results. Ph.D. thesis, Massachusetts Institute of Technology (1986)
7. Davis, F.D.: Perceived usefulness, perceived ease of use, and user acceptance of information technology. MIS Q. **13**(3), 319–340 (1989)
8. Duke, R.D.: Gaming: the Future's Language. Wiley, New York (1974)
9. Gelderman, C.J., Paul, W.T., Van Diemen, R., et al.: Choosing self-service technologies or interpersonal services? The impact of situational factors and technology-related attitudes. J. Retail. Consum. Serv. **18**(5), 414–421 (2011)
10. Greenblat, C.S.: Designing Games and Simulations: An Illustrated Handbook. Sage Publications, Inc., Newbury Park (1988)
11. Ito, G., Yamakage, S.: From kiss to tass modeling: a preliminary analysis of the segregation model incorporated with spatial data on Chicago. Jpn. J. Polit. Sci. **16**(4), 553–573 (2015)
12. Kurahashi, S.: State-of-the-art of social system research 4 model estimation and inverse simulation. J. Soc. Instrum. Control Eng. **52**(7), 588–594 (2013)
13. Liljander, V., Gillberg, F., Gummerus, J., Van Riel, A.: Technology readiness and the evaluation and adoption of self-service technologies. J. Retail. Consum. Serv. **13**(3), 177–191 (2006)
14. Meuter, M.L., Ostrom, A.L., Bitner, M.J., Roundtree, R.: The influence of technology anxiety on consumer use and experiences with self-service technologies. J. Bus. Res. **56**(11), 899–906 (2003)
15. Meuter, M.L., Ostrom, A.L., Roundtree, R.I., Bitner, M.J.: Self-service technologies: understanding customer satisfaction with technology-based service encounters. J. Market. **64**(3), 50–64 (2000)
16. Railsback, S.F., Grimm, V.: Agent-Based and Individual-Based Modeling: A Practical Introduction. Princeton University Press, Princeton (2012)
17. Rieber, L.P.: Seriously considering play: designing interactive learning environments based on the blending of microworlds, simulations, and games. Educ. Technol. Res. Dev. **44**(2), 43–58 (1996)
18. Rogers, E.M.: Diffusion of Innovations. Free Press, New York (1983)
19. Salen, K., Zimmerman, E.: Rules of Play: Game Design Fundamentals. MIT press, Cambridge (2004)
20. Sunaguchi, H., Shirai, H., Sato, R.: Evaluation of the business strategy design method using a combination of gaming and computer simulation. Stud. Simul. Gaming **26**(1), 1–8 (2016). (in Japanese)
21. Ueda, K., Kurahashi, S.: How passenger decides a check-in option in an airport. In: Social Simulation Conference (2014)
22. Ueda, K., Kurahashi, S.: The passenger decision making mechanism of self-service kiosk at the airport. In: Kurahashi, S., Ohta, Y., Arai, S., Satoh, K., Bekki, D. (eds.) JSAI-isAI 2016. LNCS (LNAI), vol. 10247, pp. 159–175. Springer, Cham (2017). https://doi.org/10.1007/978-3-319-61572-1_11
23. Ueda, K., Kurahashi, S.: Agent-based self-service technology adoption model for air-travelers: exploring best operational practices. Front. Phys. **6**, 5 (2018)
24. Watts, C., Gilbert, N.: Simulating Innovation: Computer-Based Tools for Rethinking Innovation. Edward Elgar Publishing, Northampton (2014)
25. Zaltman, G.: How Customers Think: Essential Insights into the Mind of the Market. Harvard Business Press, Boston (2003)

When We Talk About Medical Service, What Do We Concern? A Text Analysis of Weibo Data

Ke Wang[1], Chaocheng He[2], Lin Wang[1], and Jiang Wu[2(✉)]

[1] School of Information Management, Wuhan University,
Wuhan 430072, China
[2] Center for E-commerce Research and Development, Wuhan 430072, China
jiangw@whu.edu.cn

Abstract. In the information age, people are spending increasing time and energy to search and involve in health topics. And processes in social network offer new ways of investigating public opinion on this topic. However, when we are talking about medical service, what do we concern? The deterioration of doctor-patient relationship? The medical insurance policy? Or the medical achievements we have got? Understanding hot topics of medical issues and its dynamic changes help us to guide a healthy doctor-patient relationship and maintain a stable online public opinion environment. In order to figure out the question, we collected the Weibo tweets about medical information and then extracted the subjects by LDA model. Due to the feature sparsity and semantic fuzziness of short texts, this paper extended the features by using Word2vec. Finally, we summarized 14 hot subjects of medical service and analyzed the dynamic change of subjects' frequency and emotion. We find that most of the subjects are related to the medical system reform and the doctor-patient relationship. What's more, public's attitude to medical issues is gradually becoming growing positive.

Keywords: Medical service · Subject extraction · Feature extension · Weibo

1 Introduction

Concerning about themselves, their family or friends, instigated by a recent news report or scientific publication or due to general interest, people nowadays are spending increasing time and energy to search and involve in health topics, especially in the information age. Health topics such as illness and diagnosis are a social construction [1] so an exploration of social media may offer new ways of investigating public opinion on this topic. The growth of the Web 2.0 and the rising popularity of social platforms such as Twitter and Facebook have provided opportunities for people to express their opinions publicly more often than ever before [2]. Hundreds of millions of people in the United States use the Internet to find health-related information with up to 8 million people searching for health-related information on a typical day [3]. At the meantime, medical accidents have occurred frequently and the contradictions between

© Springer Nature Singapore Pte Ltd. 2018
J. Chen et al. (Eds.): KSS 2018, CCIS 949, pp. 45–57, 2018.
https://doi.org/10.1007/978-981-13-3149-7_4

doctors and patients have become increasingly severe. The reports of malignant events such as "doctors collectively kneeling down asking for apologies" and "patients' families injured anesthetist to paralysis" caused a series online public opinion outbreaks. In China the achievements in medical theory and clinical medicine are also remarkable. It is obvious that online medical public opinions have played a crucial role in social public opinions in recent years, which in turn has aroused many attentions from scholars in China.

Prior research has focused primarily on using social network to track emerging diseases [4], disseminating health-related messages [5], and understand the public's views, knowledge, attitudes, beliefs, and behaviors [6, 7]. What's more, a new area of research is rapidly developing, known as sentiment analysis, the objective of which is to translate opinions and expressions of human emotion into data that can be quantified and categorized to determine the attitude towards particular topics, services or products [8]. Sentiment analysis is currently being used in social domain [9], including public health [10]. However, there remains a blank in this research area: what are the medical public opinions? In other words, when we talk about medical service, what do we concern? It is important to find the answer in order to understand the public concerns and information needs for medical service, and further, to ease the doctor-patient contradiction and deepen the medical service reform in China. Our research objective is accordingly to figure out the main subjects and their dynamic changes of online medical service.

Social network sites such as Sina Weibo, Tencent Weibo and Twitter containing rich information about people's daily life such as childbirth, marriage and public health topics [11] have swept across the globe in recent years. While in China, Sina Weibo, a social media of news gathering, self-expression, and social participation is the main source of online public opinions. Every time a major social event happened, there will be a significant increase in the volume of blogs on Weibo. For example, after Tu Youyou got the Nobel Prize in Physiology or Medicine, her research had suddenly become a hot topic in Weibo. We therefore plan to collect Weibo tweets about medical service to do the subject extraction and its dynamic analysis. However, one thing should be noticed is that Weibo tweets are usually short texts. Traditional text mining methods are not suitable for short texts processing. Our second research objective is therefore to overcome the data sparsity of Weibo short texts by some means. In this paper, we do the subjects extraction by using LDA model. Like many other bag-of-words models, LDA model calculates the co-occurrence rules of words. But when the training corpus of texts is small or when the texts are short, the resulting distributions might be based on little evidence. Therefore, LDA is not suitable for dealing with short texts. In order to mitigate the sparseness issue of short texts, a feature extension process based on Word2vec is implemented.

The rest of this paper is organized as follows. Section 2 offers a more detailed review of online public opinion of medical service, feature extension of short texts and word embedding by Word2vec. Section 3 details methods and process. And Sect. 4 shows experiment results. Finally, in Sect. 5, we draws conclusions and introduces visions for future research.

2 Literature Review

2.1 Online Public Opinion of Medical Service

Social network dedicated to health-related topics have evolved into easily accessible participatory tools for the exchange of knowledge, experience, and opinions [11]. Internet users have been shown to be primarily interested in specific information on health problems or diseases [11–13] and in adopting a healthier lifestyle and looking for alternative points of view. In fact, the formation and the spread of online medical opinion and sentiment are related with many factors. For example, online news plays an important role in the occurrence and development of the public opinion [14–17]. Especially, medical emergency news could always cause a public opinion outbreak. Lack of properly guide of doctor-patient relationship will lead to a great damage of the hospital image and worsen the relationship between doctors and patients. Hospital may even slide into the crisis of public opinion. It's of great importance for government and social platforms to maintain a stable online public opinion environment.

2.2 Feature Extension of Short Texts

With the rapid development of Internet social media, excavating the emotional tendencies of the short text information from the Internet, the acquisition of useful information has attracted the attention of researchers [18]. However, Weibo tweets are usually short texts which is not suitable for traditional text mining methods. Compared with traditional texts, the characteristics of short texts are as follows: (a) sparseness, each short text contains very little effective information; (b) real-time, most of the internet texts are updated in real time, and the number of dynamic texts is very large; (c) irregularities, short texts are not standardized, new words are included, and noise features are numerous [19]. While feature extension is a main direction of short texts research [18]. Extending vocabulary set through external resource such as knowledge bases or using suffix tree models to construct phrases, feature extension is used to solve the problem of feature sparseness and semantic ambiguity of short texts. A research showed that a feature extension method based on frequent term sets could overcome the drawbacks of the VSM (Vector Space Model) on representing short text content [20].

2.3 Word Embedding by Word2vec

Word embedding is the collective name for a set of language modeling and feature learning techniques in NLP where words or phrases from the vocabulary are mapped to vectors of real numbers. Word2vec is one of a method to generate this mapping by neural networks [21]. Embedding vectors created by using the Word2vec algorithm have many advantages compared to earlier algorithms which calculate the co-occurrence rules of words in documents with global views. Word2vec predicts a word in its local context. In other words, Word2vec algorithm takes a large corpus of text as its input and produces a vector space, typically of several hundred dimensions, with each unique word in the corpus being assigned a corresponding vector in the space.

Word vectors are positioned in the vector space such that words that share common contexts in the corpus are located in close proximity to one another in the space. In this case, text mining will not be limited by the length of the documents. Thus Word2vec is widely used in short texts, such as classification [21], synonym extraction [22] and sentiment analysis [23]. Kusner et al. proposed a WMD (Word Mover Distance) model based on Word2vec which used the different words' vectors in two documents to measure the similarity of these two documents [24]. The classification experiment results showed that the WMD model achieved lower error rate than the Boolean model, TF-IDF, mSDA, LDA, and LSI.

3   Method

This research designs a short texts analyzing method to find out what public concern about medical service. Figure 1 shows the main process and the key techniques.

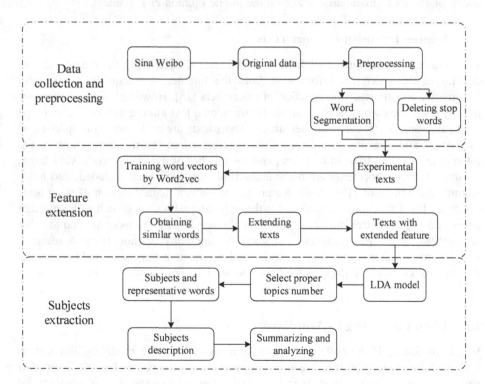

Fig. 1. Texts processing steps and technique

3.1   Training Similar Words

This paper tries to train the collected Weibo tweets by Word2vec to get the vectorial expressions for each word. Mikolov et al found that semantic and syntactic patterns can be reproduced using vector arithmetic [25]. Thus, Word2vec is also able to capture multiple different degrees of similarity between words. By calculating the cosine similarity between the word vectors, we finally get the most similar words (in aspect of semantic or syntactic) of each word. Take "新生儿/病危/被送往/医院" (means "A critically ill newborn was sent to hospital for treating") as example, Table 1 shows some similar words of the original words generated by Word2vec.

Table 1. Similar words examples.

Original word	Similar words	Cosine similarity
新生儿 （newborn）	市人民医院 （people's hospital）	0.7177569
	少儿 （children）	0.6791407
	出生 （birth）	0.66434807
病危 (critically ill)	肺炎 （pneumonia）	0.89627737
	诊室 （clinic room）	0.7956481
	家属 （families）	0.7353134
医院 (hospital)	医生 （doctor）	0.64606994
	检查 （checkup）	0.5687174
	患者 （patient）	0.5653895
救治 (treating)	检查 （checkup）	0.7374143
	家属 （families）	0.63913155
	患者 （patient）	0.63495576

We can see that the similar words obtained by word2vec are indeed related to the original words such as "新生儿(newborn)" and "出生(birth)", "救治(treating)" and "检查(checkup)", which are in line with human cognitive habits.

3.2 Feature Extension for the Texts

From Table 1 we see that the similar words of all original words in one text may be repetitious such as "families", "checkup", and "patient". Said that these three words are not the most similar word to each original word, but for the entire text, these three words appear repeatedly and provide very important semantic information for the text. So the selection of the expanded words of this text should be given priority to a certain extent of these three words.

The idea of expanding the entire text based on similar words is as follows. First, find m nearest words and the corresponding cosine similarity of each word in each text based on Word2vec. Second, iterates all similar words in one text. Find repetitious ones and calculate the sum of repetitious word's cosine similarity as the new similarity. Then we get non-repetitious similarity words set for a text. Finally, sort the similar words by their corresponding cosine similarity. Take the first n nearest words as the expanded words of the text.

Take "新生儿/病危/被送往/医院" (means "A critically ill newborn was sent to hospital for treating") as example, Fig. 2 shows the detailed procedures of feature extension method proposed by this paper. Here we set m = 3 and n = 5.

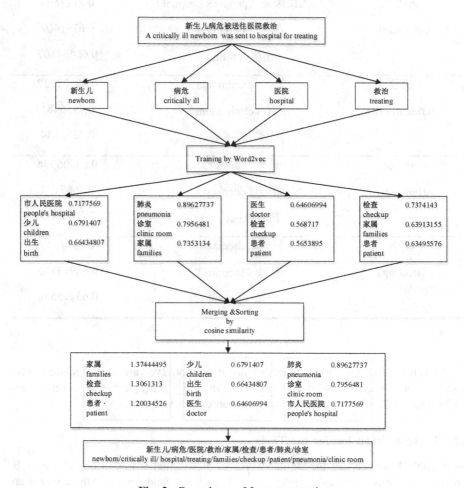

Fig. 2. Procedures of feature extension

3.3   Subject Extraction

Latent Dirichlet Allocation (LDA) is a generative statistical model that allows sets of observations to be explained by unobserved groups that explain why some parts of the data are similar. LDA is an example of a topic model and was first presented as a graphical model for topic discovery [26].

Via LDA, each medical text can be viewed as a mixture of various topics and each topic is assumed to be characterized by a particular set of words. In this research we pay more attention to the relationship of topics and the words.

4   Experiment Results

4.1   Data Collection and Preprocessing

We collected the medical related data in Weibo by Gooseeker from March 16, 2013 to March 15, 2018. We used hashtags "medical" as keywords to make queries. After eliminating approximately duplicated records this research resulted in 25939 tweets. We than created a user dictionary and stop words table to do the data preprocessing including Chinese segmentation and delete the stop words like "@", "#", "Web Link" and so on.

4.2   Parameter Settings

It is necessary to set the learning parameters of the Word2vec model. Some adjustable parameters and the settings of this research are shown in Table 2.

Table 2. Parameter of Word2vec model.

Parameter	Value	Meaning
cbow	0	Skip-Gram model
size	200	Dimensionality of the vectors is 200
window	10	Context window size is 10
hs	0	Trained with negative sampling
sample	1e-3	Threshold of Sub-sample is 0.001

Word2vec can utilize either of two model architectures to produce a distributed representation of words: continuous bag-of-words (CBOW) or continuous skip-gram. CBOW is faster while skip-gram is slower but does a better job for infrequent words. Typically, the dimensionality of the vectors used to set to be between 170 and 250 for a higher accuracy if we take skip-gram model to train the vectors. The size of the context window determines how many words before and after a given word would be included as context words of the given word. The recommended value is 10 for skip-gram and 5 for CBOW. A Word2vec model can be trained with hierarchical softmax or negative sampling. According to the authors, hierarchical softmax works better for infrequent words while negative sampling works better for frequent words and better with low dimensional vectors. High frequency words often provide little information. Words with frequency above a certain threshold may be sub-sampled to increase training speed. The sub-sampled threshold for high frequency words works the best between 1e-3 and 1e-5 [27].

In order to improve the efficiency of of LDA model training and shorten the time of feature extension. We set n, the number of extension words for each text as 25.

As for LDA topic model, the main parameters we adjusted in this research are: (a) alpha = 0.2, the bigger the alpha is, the closer each document to the same topic is. (b) beta = 0.01, the bigger beta means each topic is more focused on a few words, or each word is corresponded to one topic as far as 100% probability. (c) num_topics is the number of topics. In this research, we set the range of topics as 2 to 100. We took Perplexity as a measure of the convergence effect of LDA model training. The lower Perplexity, the better the effect.

4.3 Subject Extraction Results

Figure 3 below is the LDA model perplexities of texts with feature extension. When the topics number is 50, there is an obvious turning point in the fold line. That means in LDA model, if the topics number increase, there is no obvious decrease of perplexity.

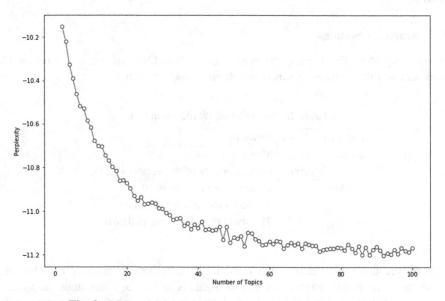

Fig. 3. LDA model perplexities of texts with feature extension

Using the most representative 10 words to express each subject, we finally summarized 14 reasonable subjects (see Table 3) when the topics number is 50.

From Table 3, we can see people's main concerns for medical service is: (a) the medical system reform including medical insurance, medical treatment in other province and the medical policy in USA. (b) The doctor-patient relationship including

Table 3. Medical service subjects.

Subject	Description	Subject	Description
1	Doctor-patient interaction	8	Medical insurance
2	Medical treatment in other province	9	TV about medical staff
3	Medical crowdfunding	10	Emergent infectious disease
4	Child medical services	11	Medical ethics
5	Medical disputes	12	Mutual understanding
6	Medical policy in USA	13	Medical achievements
7	Medical system reform	14	Media's flawed coverage

doctor-patient interaction, medical ethics, medical disputes and the false coverage of medical events. (c) Other medical related topics, such as child medical service, emergent infectious diseases, medical achievements and TV programs about medical staff.

In particular, the medical policies people discussed reflect the two aspects of medical system reform: medical insurance and medical treatment in other province. After three decades of reform, the medical system moves from the "welfare state model" to the "social insurance model". The extending of medical insurance coverage has improved the cost efficiency of Chinese hospitals. However, the current medical insurance system is still not conducive to fairness. Some scholars believe the reconstruction of medical insurance system is the general direction of medical reform in China. Medical treating in different province is also a major issue that need to be resolved. Additionally, people also care the medical policy in USA, especially the Obama Care and the critical statements by President Donald Trump about it.

The doctor-patient relationship is another hot topic of medical service. In recent years, the doctor-patient relationship in China is worsen, which manifests as increasing medical disputes and frequent medical violence. Some TV programs about medical staff inflected this problem also. Many believe the tense relationship between doctors and patients mainly results from the lack of trust. Additionally, some unscrupulous media play an important role in provoking tension between doctors and patients because of their false report. Nowadays, "effective communication", "mutual understanding" and "medical ethics" are mentioned again. People are appealing for rebuilding the trust between doctors and patients to resolve the predicament of the current doctor-patient relationship.

Moreover, medical crowdfunding is a way to raise money for critical illness from large groups of individuals. The fraudulent information and privacy problems are the main ethical concerns of crowdfunding [20]. Child medical service also has attracted people's attention. People are likely to browse the emergent infectious disease information such as H7N9 through Weibo platform because of the fast speed and wide coverage of information dissemination [21]. Finally, some medical achievements also draw the attention of public. For example, as the first Chinese woman who won the Nobel Prize, Tu Youyou has arisen lots of discussions and debates.

4.4   Evolution Analysis of Hot Topics

Understanding the dynamics of public opinion over time is especially helpful tracking the effect of medical policies and promote medical reform [28]. In order to study the changes in the hot topics of medical during 2013 and 2017, we sorted the Weibo tweets by the sum of the number of retweets, comments and agrees and manually labeled the topics of the Top 100 tweets in each year. Figure 4 displays the frequency of each topic.

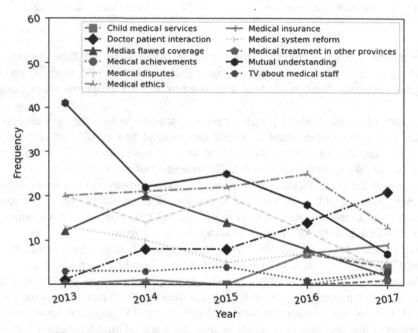

Fig. 4. Frequency of hot topics of top 100 tweets (2013–2017)

From 2013 to 2017, the main changes in the medical hot topics are as follows: (1) the number of topics has increased from 7 topics in 2013 to 11 topics in 2017. Specifically, the public's attention to the medical system reform is more concrete. Topic "medical insurance" and topic "medical treatment in other province" have appeared in 2017. Topic "child medical services" and topic "medical achievements" are also new hot topics. (2) Different medical hot topics have changed from the proportion. What is remarkable is that topic "mutual understanding" has a clear downward trend, and topic "doctor-patient interaction" has increased year by year. Analysis of the original Weibo tweets found that people's appeal for mutual understanding has gradually decreased, while the expressions of harmonious doctor-patient interaction are more common, which partly reflects the gradual warming of the relationship between doctors and patients.

Public's emotional attitude towards different topics also have changed in recent years. We did affective computing of the Top 100 tweets in each year and calculated the average sentiment scores for each topic (See Fig. 5).

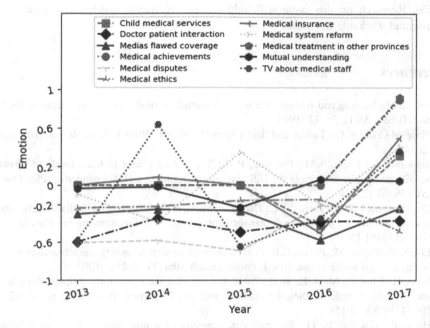

Fig. 5. Emotion change of hot topics (2013–2017)

Most of the sentiment scores in 2013 were below zero, representing more negative public sentiment towards medical topics. This situation continued until 2016. However, in 2017 the sentiment scores of each topic have been significantly improved, especially "medical insurance", "medical system reform" and "mutual understanding". As for the new topic "child medical services" and "medical treatments in other province", the sentiment were also positive. This change further proves that the peoples trust in public health has gradually recovered, and positive emotions have dominated.

5 Conclusions

This paper is written to figure out what people concern about medical service and its dynamic changes in recent years by analyzing Weibo data. We also used a feature extension method based on Word2vec to overcome data sparsity problem of short texts.

We have two major findings in this paper. First the medical system reform and the doctor-patient relationship are core points of public attentions, although the specific hot topics have some change in the last 5 years. And many other hot topics about medical service are ultimately related to these two issues. Second, public's attitude to medical issues is gradually becoming growing positive.

We plan to extend this work in two directions. First, why public's attention and attitude to medical issues have changed? A preliminary guess is that this change may be caused by some certain policies. We should take this question seriously and do more research. Second, different population groups have different concerns about medical service. Research on this issue will help the public health institutions make more specific and workable decisions.

References

1. Brown, P.: Naming and framing: the social construction of diagnosis and illness. J. Health Soc. Behav. **35**(1), 35–42 (1995)
2. Finfgeld-Connett, D.: Twitter and health science research. West J. Nurs. Res. **37**(10), 1269 (2014)
3. Signorini, A., Segre, A.M., Polgreen, P.M.: The use of twitter to track levels of disease activity and public concern in the U.S. during the influenza a H1N1 pandemic. Plos One **6** (5), e19467 (2011)
4. Charlessmith, L.E., Reynolds, T.L., Cameron, M.A., et al.: Using social media for actionable disease surveillance and outbreak management: a systematic literature review. Plos One **10** (10), e139701 (2015)
5. Lister, C., Royne, M., Payne, H.E., et al.: The laugh model: reframing and rebranding public health through social media. Am. J. Public Health **105**(11), 2245 (2015)
6. Woo, H., Cho, Y., Shim, E., et al.: Public trauma after the sewol ferry disaster: the role of social media in understanding the public mood. Int. J. Environ. Res. Public Health **12**(9), 10974–10983 (2015)
7. Hays, R., Dakerwhite, G.: The care data consensus? a qualitative analysis of opinions expressed on twitter. BMC Public Health **15**(1), 838 (2015)
8. Balahur, A., Mihalcea, R., Montoyo, A.: Computational approaches to subjectivity and sentiment analysis: present and envisaged methods and applications. Comput. Speech Lang. **28**(1), 1–6 (2014)
9. Liu, B.: Sentiment Analysis and Opinion Mining. Morgan & Claypool Publishers, San Rafael (2012)
10. Ji, X., Chun, S.A., Wei, Z., et al.: Twitter sentiment classification for measuring public health concerns. Soc. Netw. Anal. Min. **5**(1), 13 (2015)
11. Choudhury, M.D., Counts, S., Horvitz, E.: Major life changes and behavioral markers in social media: case of childbirth. In: Conference on Computer Supported Cooperative Work, pp. 1431–1442. ACM (2013)
12. Ybarra, M.L.: Help seeking behavior and the internet: a national survey. Int. J. Med. Informatics **75**(1), 29–41 (2006)
13. Rice, R.E.: Influences, usage, and outcomes of internet health information searching: multivariate results from the pew surveys. Int. J. Med. Informatics **75**(1), 8–28 (2006)
14. Kagashe, I., Yan, Z., Suheryani, I.: Enhancing seasonal influenza surveillance: topic analysis of widely used medicinal drugs using twitter data. J. Med. Internet Res. **19**(9), e315 (2017)
15. Didi, S., Quoc, N.D., Georgina, K., et al.: Characterizing twitter discussions about HPV vaccines using topic modeling and community detection. J. Med. Internet Res. **18**(8), e232 (2016)
16. Greaves, F., Ramirezcano, D., Millett, C., et al.: Use of sentiment analysis for capturing patient experience from free-text comments posted online. J. Med. Internet Res. **15**(11), e239 (2013)

17. Metwally, O., Blumberg, S., Ladabaum, U., et al.: Using social media to characterize public sentiment toward medical interventions commonly used for cancer screening: an observational study. J. Med. Internet Res. **19**(6), e200 (2017)
18. Liu, Y., Zhu, X.: Short text sentiment classification based on feature extension and ensemble classifier. In: International Conference on Computer-Aided Design, Manufacturing Modeling and Simulation, pp. 020051 (2018)
19. Ji, Y.L., Dernoncourt, F.: Sequential Short-Text Classification With Recurrent and Convolutional Neural Networks, pp. 515–520 (2016)
20. Man, Y.: Feature extension for short text categorization using frequent term sets. Procedia Comput. Sci. **31**, 663–670 (2014)
21. Liu, W., Cao, Z., Wang, J., et al.: Short text classification based on Wikipedia and Word2vec. In: IEEE International Conference on Computer and Communications, pp. 1195–1200. IEEE (2017)
22. Zhang, L., Li, J., Wang, C.: Automatic synonym extraction using Word2Vec and spectral clustering. In: Chinese Control Conference, pp. 5629–5632 (2017)
23. Yu, L.C., Wang, J., Lai, K.R., et al.: Refining word embeddings using intensity scores for sentiment analysis. IEEE/ACM Trans. Audio Speech Lang. Process. **26**(3), 671–681 (2018)
24. Kusner, M.J., Sun, Y., Kolkin, N.I., et al.: From word embeddings to document distances. In: International Conference on International Conference on Machine Learning, JMLR.org, pp. 957–966 (2015)
25. Blei, D.M., Ng, A.Y., Jordan, M.I.: Latent dirichlet allocation. J. Mach. Learn. Res. **3**, 993–1022 (2003)
26. Maaten, L.V.D., Hinton, G.: Visualizing data using t-SNE. J. Mach. Learn. Res. **9**(2605), 2579–2605 (2008)
27. Chen, Z., Barros, C.P., Hou, X.: Has the medical reform improved the cost efficiency of Chinese hospitals? Soc. Sci. J. **53**(4), 510–520 (2016)
28. Jacobs, L.R., Mettler, S.: Why public opinion changes: the implications for health and health policy. J. Health Polit. Policy Law **36**(6), 917–933 (2011)

Modeling Wicked Problems in Healthcare Using Interactive Qualitative Analysis: The Case of Patients' Internet Usage

Renuka Devi S. Karthikeyan, Prakash Sai Lokachari[✉],
and Nargis Pervin

Department of Management Studies, Indian Institute of Technology Madras,
Chennai, India
lps@iitm.ac.in

Abstract. Wicked problems are embedded with attributes such as lack of understanding, multiple stakeholders' involvement in solution implementation and the lack of opportunity of undoing the solution implemented. Hence, a robust methodology is required to understand its nature. Interactive Qualitative Analysis (IQA) is a systems method that caters to the need of understanding the phenomenon while also provisioning means to understand different stakeholders' perceptions of the phenomenon. This study demonstrates the use of IQA in understanding a wicked problem in healthcare sector dealing with patients' internet usage. Our analysis reveals that patients elicit the need of the internet in three new realms, namely, hospital choice, physician choice, and online support services, which were not apparent in previous studies on the same context. The study's findings provide insights that could lead to development of strategies for meta-services within healthcare sector involving information dissemination for patients through the internet.

Keywords: Wicked problems · Interactive Qualitative Analysis
Patients' internet usage · Healthcare

1 Introduction

Internet being one of the disruptive innovations is expected to beat the decentralization of medical knowledge, which has largely been confined within the professionals in the healthcare sector. It has the potential to promote medical pluralism and serve as a platform for providing information on alternate treatment options. In fact, it advocates patient empowerment, which could eventually lead to a patient-centered healthcare system [3]. Though online health information was considered only secondary to the opinions of healthcare professionals, they are still sought by a majority of the patients, as and when additional information not provided by physicians, is needed. Such practices affect the patients' decision in treatment adherence and physician choice [2, 4, 26, 27]. Consequently, a systematic investigation of patients' internet usage and its influence on decision making is of paramount importance. However, one must note that online sources still hold the possibility of disseminating inaccurate and unreliable

© Springer Nature Singapore Pte Ltd. 2018
J. Chen et al. (Eds.): KSS 2018, CCIS 949, pp. 58–70, 2018.
https://doi.org/10.1007/978-981-13-3149-7_5

health information with potentials of drastic outcomes, and hence, dealing with patient's internet usage is not straightforward.

Wicked problems, also known as social messes, are unstructured problems that cannot be mapped based on a cause-effect relationship and hence are difficult to model. Attempting a solution often aggravates or changes its nature. In this context, the case of patients' internet usage is an alluring problem. Although the online healthcare related information is inadequate and also the healthcare system is reluctant to acknowledge this fact, in the current scenario, the use of internet among the patient community is growing exponentially. In prior literature, typically patients' internet usage scenario has been studied under a Communications perspective [14]. Thus, we anticipate that classifying this as a wicked problem to gain a holistic understanding beyond a communication theoretical lens is of utmost importance, which requires a second generation systems approach, wherein the participants are able to visualize the problem and solution through an iterative process of arguments and inferences [25]. Head and Alford have considered that the knowledge-power differential poses as a challenge in addressing wicked problems [11], which can be dealt with Interactive Qualitative Analysis (IQA). IQA considers both power and knowledge to be largely interdependent in understanding a phenomenon. It is a systems method of qualitative investigation, which could help in viewing a phenomenon from the perspectives of different constituencies. A constituency is a group of people who experience a phenomenon similarly and one constituency differs from another based on their proximity to or power over a phenomenon. It helps capture a conceptual mind map of the problem from the participants' viewpoint and also has the inherent quality of minimizing researcher bias with increased participant contribution in the exploration. Considering the power of IQA in modeling a wicked problem, we have employed the method in comprehending patients' internet usage and deriving the drivers and outcomes of the phenomenon [20].

2 Research Background

2.1 Internet Usage Among Patients

Internet usage primarily serves as a source of information to bridge the gap between the patients and the physicians and has also increased their participation in the service delivery. The number of internet users had kept progressively increasing with the passage of years [12, 22, 26] and it has been noted that patients belonging to younger generation and those patients with higher educational qualification, higher income and females are prolific users of the internet [2]. While two decades ago, patients cited lack of interest, expensiveness, inability to "know where to start" and a general reluctance towards online mode of communication as the reasons for not using the internet, its usage has grown by 1,052% in this time span [27, 28]. It is a preferred source of information in healthcare domain as it is easily accessible and the information helps in improving the confidence in the treatment [15]. However, patients have recounted certain difficulties in searching for the appropriate information due to information overload, difficulty in comprehending the information and inability to check for accuracy [24].

Patients seek online information to get additional information on their health condition, diagnosis and treatment [15, 21, 24], information on dietary and exercise habits [24] and to validate information provided by the physician [21]. The internet usage tends to increase if they are recently diagnosed with diseases or if they have been affected by the disease for a longer duration with their condition worsening. Patients with higher disability tend to use the internet more as a source providing interaction with specialists and support groups [15]. However, patients hesitate to consult with their physicians on the health related information found online; while on the other hand, physicians are also less enthusiastic in discussing online health information with their patients, simply because it lengthens the duration of the consultation and also leads to unnecessary investigation [15, 22]. But as this leads to failure in verifying the reliability of the content, it is also vital that physicians encourage patients to discuss their internet usage habits during consultation and guide them towards reliable sites [1, 15, 21, 24, 26].

Patients also have the opportunity to share their experiences online, rate and recommend their physicians. Contrary to popular belief that patients tend to go online to record their negative experiences, patients in UK were found to have given around 64% positive responses (recommending a particular physician). The recommendations were positively related with the provision of care received and not related to the clinical outcomes [7]. This reinstates that physician's behavior during consultation has an important role to play in the overall perception of the patients.

2.2 Wicked Problems

Problems are loosely classified into two: tame problems and wicked problems. Tame problems can be characterized, understood and can be solved using a stopping criteria. On the other hand, wicked problems are those that cannot have definitive formulation. The understanding of the problem stems from trying to arrive at all the possible solutions. Attempting a single solution doesn't necessarily solve the problem rather morphs it into a different condition and hence lacking a stopping point. Every attempt to solve a wicked problem requires significant effort and there is no scope for reverting back the solution. Planning problems are usually considered wicked problems because an attempt at a solution cuts across all levels of authority and requires the participation of various departments. Also the attempted solutions cannot be undone and thus, planners have no "right to be wrong" [25].

While there are arguments that problems cannot be easily distinguished into wicked and tame problems, Head and Alford have suggested looking at wicked problems by disintegrating them into characteristics and understanding why it is difficult to understand or solve them. They advocate the presence of 'degrees of wickedness' and propose to seek solutions by considering the structural and political authority of implementation and arrive at makeshift control over the problems [11].

Internet Usage as a Wicked Problem

Patients' internet usage usually stems from the aspects that either they are not financially endowed for access to quality healthcare services (lack of insurance coverage) or they are unable to receive services due to the circumstances or they are not satisfied with the quality of healthcare services that they are currently receiving which serves as a substitute for self-diagnosis and treatment [18]. It was also commonly noted in many studies that patients rarely discuss their online information seeking habits either not feeling the need to do it or for fear that this might be construed as a challenge of the physician's competency [10].

A patient-centric approach could promote a symbiotic relationship between the patient and the physician: the patients gather in-depth information of the clinical condition they possess and the physician, with their knowledge, is able to guide the patient on its reliability while also having the opportunity to gain any additional information or insights based on the patients reading [16]. However, the lack of opportunity to increase the consultation time is a drawback. A physician opting to take a defensive stand, trying to impose their medical authority and steering the consultation towards their choice of treatment stops the patient from disclosing any information that he/she may have gained online [9, 16]. In such circumstances, to compensate for lack of physician participation, health literacy turns out as a required factor to identify accurate medical information [8]. Hence, promoting health literacy is also important. From the perspective of internet usage by patients, the understanding of the problem has moved from physician participation to improvement in health literacy. Health literacy improvement can be achieved by implementation of inclusive educational curriculum to make people versatile with medical information. However, if we consider the financial deficiency perspective leading to internet usage, it goes into the realms of National Health policy changes that are required to make healthcare services affordable for all classes of citizens. From the above discussion, it is apparent that the phenomenon of internet usage by the patients branches out to multiple dimensions and requires involvement of several stakeholders. Thus, to gain a holistic view of the problem, we need to conceptualize it from a wicked problem perspective.

To gain better understanding of the issue, we require a systems approach wherein all levels of authority over a decision are taken into consideration. Majority of the extant studies on patients' internet usage have dived into a quantitative analysis of the situation, using questionnaires, without a social understanding or theory supporting it, which demands the need for a qualitative inquisition to cull out the underlying themes governing the phenomenon. In particular, the wicked problem-based approach of understanding the problem solicits the need for a systems approach. Therefore, the use of the Interactive Qualitative Analysis, a systems method, turns out to be a convenient alternative to regular qualitative inquiry for the research problem, which has been discussed in the next section.

The objectives of this research are as follows:

- To understand the perceptions of patients, explore influential factors, and expected outcomes concerning the usage of the Internet during illness.
- To explore the opportunities of examining and modeling Wicked Problems through Interactive Qualitative Analysis.

3 Methodology

3.1 Interactive Qualitative Analysis

Interactive Qualitative Analysis (IQA) is a qualitative research methodology built using the principles of systems theory. The IQA process categorizes the participants (referred as constituencies) based on their proximity to and influence over the phenomenon. Northcutt and McCoy [20] have defined constituency as "a group of people who have shared understanding of the phenomenon". The methodology involves a modified form of focus group discourse and interviews. The inculcation of both these investigation methods helps in understanding the System reality both from the individual and the group perspectives.

IQA as a method is founded on both inductive and deductive reasoning. The elements of the systems are called as affinities. The participants, first cull out the affinities, and refine their meanings (inductive approach) followed by establishing their interrelationships (deductive approach) [20]. This gives a profound understanding of the system and also minimizes researcher's bias. The outcome of an IQA study is the System Influence Diagram (SID), constructed with the interrelationships between the affinities. The SID is a representation of the mind map of the constituencies regarding the phenomenon under study and is characterized by the presence of the affinities portrayed as drivers and outcomes [20].

Process Flow of IQA

The IQA process involves two phases: IQA focus group discussion followed by IQA interview. The process revolves around a single question. The first step of the focus group discussion is individualistic where each participant thinks over the question for some time and then jots down all the thoughts, which we call as the brainstorming session. Then, in the inductive coding phase, the write-ups are grouped together under various categories by the participants with the help of the researcher. These categories are referred to as affinities in an IQA study. The researcher then starts an open discussion with all the participants towards naming each of the affinities.

The participants are then teamed up in a group of two (dyads) and are requested to provide a definition for the affinities and also check for inter-relationships with other affinities by building an affinity relationship table. The affinity relationship table is constructed by the participants by considering the pairwise relationships between the affinities. Considering two affinities A and B, the participants mention A in the cell if affinity A influences B, or B if B influences A. The participants leave the cell blank if they feel there is no relationship between A and B. Table 1 describes each of these five steps in the IQA process and the last column lists the time taken for each of these activities in minutes. The affinity relationship table paves the way for developing a System Influence Diagram that is characterized by the presence of affinities divided into drivers and outcomes. As the IQA question serves as the initiator of the whole process, it is pivotal to arrive at the right question regarding the phenomenon under study. For our current study, the IQA question is "What is the role of Internet in addressing your needs during illness?".

Table 1. Process of interactive qualitative analysis focus group for the current study

Activity	Description	Participation type	Time (in mins)
Brainstorming session	Participants write about all the experiences and opinions concerning the question on the cards provided to them	Individual	20
Inductive coding	Participants reassess all the cards and group similar cards together	Individual	20
Axial coding	Participants partake in clarifying, reorganizing and naming the affinities (by grouping the cards)	Group	10
Defining affinity	Participants arrive at a definition for the affinities by considering, comparing, and contrasting the details and enrich the affinity understanding by citing experiences and examples	Dyad	15
Constructing affinity relationship table	Participants consider the pairwise relationship between affinities and specify which affinity influences the other	Dyad	15

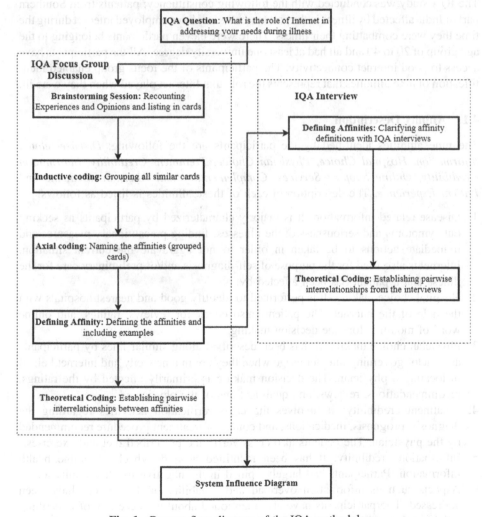

Fig. 1. Process flow diagram of the IQA methodology

The second phase of the process are the IQA interviews, where the participants are provided with an explanation of the affinities provided by the focus group participants and are then asked to recount their understanding and/or experiences pertaining to the affinities. The order in which the affinities are presented to the participants is shuffled to minimize order bias. The researcher then guides the interviewee to extract the inter-relationships between the affinities and a pairwise comparison is also derived. This provides an understanding of the individual reality of the system and also helps in clarifying the meaning of each of the affinities. Figure 1 depicts a summary of the process flow of the methodology.

4 Research Findings

The IQA study was conducted with the following constituency: patients from Southern part of India affected by illness in the past 6 months and have employed internet during the time they were combatting their illness. There were eleven participants belonging to the age group of 20 to 40 and all had at least one university degree. All the participants have access to good internet connectivity. The participants of the focus group enabled identification of nine affinities that represents the role that Internet played when they were ill.

4.1 Affinity Description

The nine affinities identified by the participants are the following: *Disease-related Information, Hospital Choice, Physician Choice, Treatment Credibility, Information Credibility, Online Support Services, Complementary and Alternate Medicine* and *Patient Experience*. The description of each of these affinities is listed as follows:

1. Disease related information: It is majorly characterized by participants as seeking out symptoms and seriousness of the illnesses; finding precautionary measures and immediate actions to be taken in order to not aggravate the current situation. Internet is also used for the purpose of self-diagnosis and/or preliminary care for the illness that a patient has been affected by.
2. Hospital choice making: It is performed to identify good and nearest hospitals with the help of the internet. The patients also check for reviews, ratings, and online word of mouth before the decision making.
3. Physician choice making: It has been described along similar lines by participants as a factor governing internet usage when they are in a new city and internet helped in locating a physician. The decision making is primarily affected by the ratings, recommendations, reviews, and qualifications of the physicians.
4. Treatment credibility: It involves the cross-verification activities regarding the diagnosis, prognosis, medications, and course of treatment procedure recommended by the physician. The patients also cross-verify the prescribed diet and exercises.
5. Information credibility: It has been identified as a drawback of online health information. Participants had largely opined in the negative for this affinity alone. Aspects such as information overload and reliability of the sources have been addressed. The participants have also mentioned about the necessity of unearthing reliable information by looking for journals or credible sources of information.

6. Emotional support: Internet has been realized as a means of staying in touch with family and friends to build moral and emotional strength in oneself and also provides an opportunity to connect with people online who are affected by the same condition.
7. Online support services: Internet facilitates to book appointments online for consultation, diagnostics, and home tests, order medicines and prescribed dietary foods, etc.
8. Complementary and Alternate Medicine: It is one of the most richly explained motives behind usage of internet during illness. Natural or home remedies, alternative therapies such as Ayurveda, Homeopathy, or options such as yoga and exercises are considered as alternatives to allopathic medication and treatment.
9. Patient Experience: It consists of treatment experience, reviews in social media and specific online portals (like Practo in South East Asia). The participants had expressed as to how providing reviews and ratings were reasons behind going online, so that their inputs can help in other patients' decision making in the future.

The participants have also explored the pairwise relationship of the affinities and each dyad provided an interrelationship table. The cells are filled with the affinity that influences the other affinity, for example, if we consider affinity 1 and 2, value 1 in cell implies that affinity 1 leads affinity 2. The consolidated table (as given in Table 2) is derived from the affinity relationship table provided by the individual dyads. Now to consolidate the affinity values by the five groups, the following procedure has been carried out. Considering the relationship between affinity 1 and 2, if 3 teams had recommended that affinity 1 influences affinity 2 while 2 teams recommended 2 influences 1, the consolidated table will consist the value 1 in the intersecting cell, indicating that the affinity 1 is considered to be influential by a majority vote.

The 'Out' column indicates the number of relationships in which the particular affinity influences the other affinities and the 'In' column indicates the number of relationships in which the affinity in question is influenced by other affinities. The 'Del' column lists the delta value (difference between the 'outs' and 'ins' of an affinity) which determines whether it is a driver or an outcome. Positive delta value indicates that the affinity acts as a driver in the system and negative delta value indicates that the affinity acts as an outcome in the system.

Table 2. Affinity interrelationship table

	1	2	3	4	5	6	7	8	9	Out	In	Del
1	–	1	1	1	1		1	1		6	0	6
2	1	–	3		5			8	2	1	4	–3
3	1	3	–	3	5	6		8	3	3	4	–1
4	1		3	–	5	6		8	9	0	6	–6
5	1	5	5	5	–	5	7	5		5	2	3
6			6	6	5	–		6		3	1	2
7	1			7		–		1		1	1	0
8	1	8	8	8	5	6		–	8	4	3	1
9		2	3	9				8	–	1	3	–2

4.2　System Influence Diagram

All the pairwise relationships have been grouped in descending order based on the number of votes. Employing the recommended Pareto Principle by the authors [20], those pairwise relationships that account for 80% of the System explanation, was considered for developing a Cluttered System Influence Diagram (SID) as described in Fig. 2.

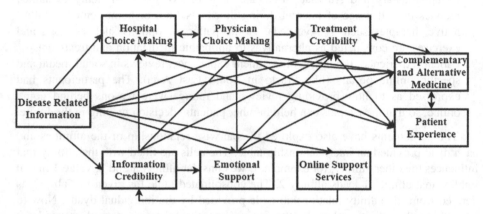

Fig. 2. Cluttered system influence diagram

The Cluttered SID is uncluttered by considering the redundant links. For example, consider the affinities: Disease related information, Information Credibility and Online Support Services. Disease related information influences both Information Credibility and Online Support Service whereas Online Support Services influence Information Credibility (please note the direction of the arrow in Fig. 2). The link between Disease Related information and Information Credibility is removed as it is already explained by the presence of Online Support Services. This redundant link removal is carried out while keeping in mind to retain those relationships that hold higher votes. The final uncluttered SID is represented in Fig. 3. The delta values from the interrelationship table are sorted in descending order and the corresponding affinities are segregated as drivers and outcomes. In general, the affinities with positive values are considered as drivers, those with negative values as outcomes, and '0' as pivot as prescribed by Northcutt and McCoy [20]. The affinities are in general then clustered into primary and secondary drivers based on higher and lower positive delta values, respectively. Similarly, affinities are clustered as primary and secondary outcomes based on higher and lower negative values, respectively. The pivot affinity is generally influenced by all the drivers and also acts as an influencer for all the outcomes. However, we have considered all affinities with non-negative values including '0' as drivers and the rest as the outcomes. While all the other affinities have been clustered as recommended, Online support services is retained as a primary driver as it does not have any relationship with any other affinities. All the primary and secondary drivers and outcomes have been depicted in Fig. 3. Considering the affinity description provided by the participants and the uncluttered SID, it has been observed that the primary drivers are those affinities that help in self-diagnosis and the secondary drivers are the ones that serve as means to self-treatment. The outcomes are those affinities pertaining to pre-consultation and post-consultation activities.

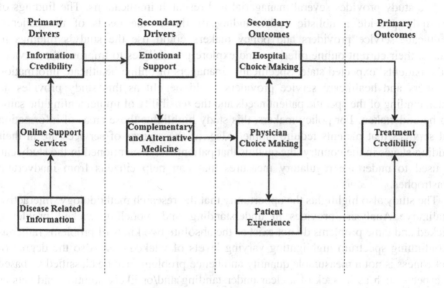

Fig. 3. Uncluttered system influence diagram

5 Discussion

This study unraveled important aspects that govern the usage of internet, which is not immediately apparent in extant literature unless the research was particularly addressing the affinity. The affinities namely, Hospital Choice Making, Physician Choice making and Online Support Services, emerged along with other six affinities through IQA process. In the case of hospital and physician choice, literature was largely based on eliciting the factors governing the decision and not the means towards identifying information concerning those factors [6, 13, 19]. This study reports internet as one of the sources. Also the previous studies have highlighted that the availability of online content depicting patients' experiences have usually been utilized for pre-consultation decision making. However, that aspect has already been accounted for in hospital and physician choice making. Our study has also indicated that patients go online to provide reviews and ratings to help other patients in their decision making. We also found a new affinity named Online Support Services that covers activities such as online appointment booking, online medicine purchase, online food purchase, etc. While these have been studied individually as process improvement opportunities (online appointment booking) or as studies on dietary/usage habits and their adverse effects (online food and medicines purchase) [5, 23], they have not been consolidated under a single concept until now. These results highlight the potent of the methodology in unraveling hidden affinities which were not apparent in extant studies. The technique's robustness has also been highlighted in the aspect that the outcome of the IQA process, the SID has provided a procedural use of internet by patients moving from information gathering for self-diagnosis to self-treatment and then moving onto activities that happen before and after consultation.

The study provides several managerial and research implications. The findings of the study provide a holistic understanding of the online needs of the patients. Healthcare service providers and policy makers could use the study's findings to analyze their current online offerings and explore possibilities to improve the services to the patients' expected state. Specifically, managers of online healthcare information providers and healthcare service providers could benefit as this study provides an understanding of the specific patient needs and the feasibility of implementing the same can be undertaken. For policy makers, this study highlights those areas of information and services that patients require online. Also, as these areas of services are in their budding stages in the country, the qualms that patients have mentioned in the study can be used to undertake regulatory measures that can help citizens from inadvertent catastrophes.

The study also highlights the opportunity that the research methodology, Interactive Qualitative Analysis, provides in understanding and modeling wicked problems. Wicked and tame problems do not exist as the absolute two kinds of problems rather as a continuing spectrum highlighting varying levels of wickedness. Also the degree of wickedness is not a measurable quantity and hence problems can be classified so based on aspects such as its lack of a clear understanding and/or likely solution and acts as Type-2 problem situation in itself [11]. IQA as a method takes into account the importance of both knowledge and power in understanding a phenomenon. The phenomenon considered for the study- Patients' Internet Usage, involves multiple stakeholders namely, patients, caregivers, physicians and healthcare professionals, information and service providers, and policy makers. This study has considered the constituency, patients, who are closest to the phenomenon but lack the power to account for issues such as reliability, online services, etc. The IQA process can be executed similarly for the same context with the other constituencies (stakeholders) as well for such online services. Such an extension to the current study would then provide a mind map of their perceptions of the phenomenon, highlight the gaps in their mind maps and also provide insights on the realistic options of implementation. It also paves way for constructive dialogue and collaboration to happen among various stakeholders as it provides a mutual understanding of each other's capacity and misgivings which has been recommended as one of the strategies to deal with wicked problems by Head and Alford [11]. Hence, IQA dominates over other methodologies in modeling and understanding wicked problems in general.

The study is not without limitations. The findings may be subjected to recall bias as it was a retrospective investigation. Also, the constituency, based on which the study was conducted, consisted of those citizens who had access to good internet connection, latest technology and communication devices and are also well-versed in using latest gadgets. We have considered Indian patients as participants of the study which limits the generalizability of our findings. Although the internet coverage in India is expected to reach around 500 million by June, 2018 [17] as per The Internet and Mobile

Association of India, the distribution is not uniformly spread across the geography. Those participants with access to resources have been considered for the study as that gives us insights into what could be considered as features required by both regular and novice users.

As a continuation to this study, we are planning to conduct a quantitative investigation using a questionnaire survey to understand the degree of usage of internet by patients with respect to the mentioned affinities and also uncover areas that need attention from the management which could be used for strategic positioning of players in the healthcare service sector.

References

1. Akerkar, S.M., Kanitkar, M., Bichile, L.S.: Use of the Internet as a resource of health information by patients: a clinic-based study in the Indian population. J. Postgrad. Med. **51**(2), 116 (2005)
2. AlGhamdi, K.M., Moussa, N.A.: Internet use by the public to search for health-related information. Int. J. Med. Inform. **81**(6), 363–373 (2012). https://doi.org/10.1016/j.ijmedinf.2011.12.004
3. Broom, A., Tovey, P.: The role of the Internet in cancer patients' engagement with complementary and alternative treatments. Health **12**(2), 139–155 (2008). https://doi.org/10.1177/1363459307086841
4. Castleton, K., Fong, T., Wang-Gillam, A., et al.: A survey of Internet utilization among patients with cancer. Support. Care Cancer **19**(8), 1183–1190 (2011). https://doi.org/10.1007/s00520-010-0935-5
5. Dusheiko, M., Gravelle, H.: Choosing and booking—and attending? impact of an electronic booking system on outpatient referrals and non-attendances. Health Econ. **27**(2), 357–371 (2018). https://doi.org/10.1002/hec.3552
6. Emmert, M., Meier, F., Pisch, F., Sander, U.: Physician choice making and characteristics associated with using physician-rating websites: cross-sectional study. J. Med. Internet Res. **15**(8) (2013). https://doi.org/10.2196/jmir.2702
7. Greaves, F., et al.: Patients' ratings of family physician practices on the internet: usage and associations with conventional measures of quality in the English National Health Service. J. Med. Internet Res. **14**(5) (2012). https://doi.org/10.2196/jmir.2280
8. Gutierrez, N., Kindratt, T.B., Pagels, P., Foster, B., Gimpel, N.E.: Health literacy, health information seeking behaviors and internet use among patients attending a private and public clinic in the same geographic area. J. Community Health **39**(1), 83–89 (2014). https://doi.org/10.1007/s10900-013-9742-5
9. Hart, A., Henwood, F., Wyatt, S.: The role of the Internet in patient-practitioner relationships: findings from a qualitative research study. J. Med. Internet Res. **6**(3) (2004). https://doi.org/10.2196/jmir.6.3.e36
10. Hay, M.C., Cadigan, R.J., et al.: Prepared patients: internet information seeking by new rheumatology patients. Arthritis Care Res. **59**(4), 575–582 (2008). https://doi.org/10.1002/art.23533
11. Head, B.W., Alford, J.: Wicked problems: implications for public policy and management. Adm. Soc. **47**(6), 711–739 (2015). https://doi.org/10.1177/0095399713481601

12. Jadad, A.R., Sigouin, C., Cocking, L., Booker, L., Whelan, T., Browman, G.: Internet use among physicians, nurses, and their patients. JAMA 286(12), 1451–1452 (2001). https://doi.org/10.1001/jama.286.12.1447

13. Lane, P.M., Lindquist, J.D.: Hospital choice: a summary of the key empirical and hypothetical findings of the 1980s. Mark. Health Serv. 8(4), 5 (1988). PMID: 10303067

14. Lee, S.Y., Hawkins, R.: Why do patients seek an alternative channel? the effects of unmet needs on patients' health-related Internet use. J. Health Commun. 15(2), 152–166 (2010). https://doi.org/10.1080/10810730903528033

15. Lejbkowicz, I., Paperna, T., Stein, N., Dishon, S., Miller, A.: Internet usage by patients with multiple sclerosis: implications to participatory medicine and personalized healthcare. Multiple sclerosis international (2010). https://doi.org/10.1155/2010/640749

16. McMullan, M.: Patients using the Internet to obtain health information: how this affects the patient–health professional relationship. Patient Educ. Couns. 63(1–2), 24–28 (2006). https://doi.org/10.1016/j.pec.2005.10.006

17. Media Inner: IAMAI. http://www.iamai.in/media/details/4990

18. Millard, R.W., Fintak, P.A.: Use of the Internet by patients with chronic illness. Dis. Manag. Health Outcomes 10(3), 187–194 (2002). https://doi.org/10.2165/00115677-200210030-00006

19. Niehues, S.M., Emmert, M., Haas, M., Schöffski, O., Hamm, B.: The impact of the emergence of internet hospital rating sites on patients' choice: a quality evaluation and examination of the patterns of approach. Int. J. Technol. Mark. 7(1), 4–19 (2012). https://doi.org/10.1504/IJTMKT.2012.046435

20. Northcutt, N., McCoy, D.: Interactive Qualitative Analysis: A Systems Method for Qualitative Research. SAGE Publications, London (2004)

21. Pereira, J.L., Koski, S., Hanson, J., Bruera, E.D., Mackey, J.R.: Internet usage among women with breast cancer: an exploratory study. Clin. Breast Cancer 1(2), 148–153 (2000). https://doi.org/10.3816/CBC.2000.n.013

22. Potts, H.W., Wyatt, J.C.: Survey of doctors' experience of patients using the Internet. J. Med. Internet Res. 4(1) (2002). https://doi.org/10.2196/jmir.4.1.e5

23. Rahkovsky, I., Anekwe, T., Gregory, C.: Chronic disease, prescription medications, and food purchases. Am. J. Health Promot. 32(4), 916–924 (2018). https://doi.org/10.1177/0890117117740935

24. Rice, R.E.: Influences, usage, and outcomes of Internet health information searching: multivariate results from the Pew surveys. Int. J. Med. Inform. 75(1), 8–28 (2006). https://doi.org/10.1016/j.ijmedinf.2005.07.032

25. Rittel, H.W., Webber, M.M.: Dilemmas in a general theory of planning. Policy Sci. 4(2), 155–169 (1973). https://doi.org/10.1007/BF01405730

26. Trotter, M.I., Morgan, D.W.: Patients' use of the Internet for health related matters: a study of Internet usage in 2000 and 2006. Health Inform. J. 14(3), 175–181 (2008). https://doi.org/10.1177/1081180X08092828

27. Välimäki, M., Nenonen, H., Koivunen, M., Suhonen, R.: Patients' perceptions of Internet usage and their opportunity to obtain health information. Med. Inform. Internet Med. 32(4), 305–314 (2007). https://doi.org/10.1080/14639230701819792

28. World Internet Users and 2018 Population Stats. https://www.internetworldstats.com/stats.htm

Mining Typical Drug Use Patterns Based on Patient Similarity from Electronic Medical Records

Jingfeng Chen[1], Chonghui Guo[1(✉)], Leilei Sun[2], and Menglin Lu[1]

[1] Institute of Systems Engineering, Dalian University of Technology,
Dalian, China
cjfeng2015@mail.dlut.edu.cn, dlutguo@dlut.edu.cn
[2] School of Economics and Management, Tsinghua University, Beijing, China
sunll@sem.tsinghua.edu.cn

Abstract. Drug use is an important part of patient treatment process to cure and prevent disease, following the strict application guidelines of clinical drugs. The availability of free and massive patient electronic medical records (EMRs) provides a new chance to mine drug use patterns by designing automatic discovery methods. In this paper, we propose a data-driven method to mine typical drug use patterns from EMRs. Firstly, we use a set of quintuple to define drug use distribution feature (*DUDF*) for each drug and represent patient treatment record with *DUDF* vector (*DUDFV*). Then we design a similarity measure method to compute the similarity between pairwise patient treatment records. Next we adopt affinity propagation (AP) clustering algorithm to cluster all patient treatment records, extract typical drug use patterns including typical drug use set, typical drug use day set, and the *DUDF* of each typical drug, and further evaluate and label typical drug use patterns with demographic and diagnostic information. Finally, experimental results on a real-world EMR data of sepsis patients show that our approach can effectively extract typical drug use patterns and develop standard treatments for patients based on their demographic and diagnostic information.

Keywords: Medical data miming · Typical drug use patterns
Similarity measure · Electronic medical records

1 Introduction

The rational drug use requires that "patients receive medications appropriate to their clinical needs, in doses that meet their own individual requirements for an adequate period of time, at the lowest cost to them and their community" [1, 2]. And for curing and preventing a specific disease, drug use is a crucial part of patient treatment process relying on standard treatment guidelines and essential drug lists. However, in clinical practice, drug use involves a series of medical problems, such as patient condition admitted to hospital, age and gender, disease type and severity, complications, drug combination and type, the drug dose per day, the start and end time of drug, the lasting time of drug, and so on, thus formulating such a standard drug use guideline, especially

© Springer Nature Singapore Pte Ltd. 2018
J. Chen et al. (Eds.): KSS 2018, CCIS 949, pp. 71–86, 2018.
https://doi.org/10.1007/978-981-13-3149-7_6

for a new disease, is very labor-intensive and time-consuming. Recently, with the development of electronic healthcare information systems, more and more electronic medical record (EMR) data is collected from hospitals and ready to be analyzed [3]. EMRs conceal an untapped reservoir of knowledge about the specific way of drug therapy activities performed on the similar patients [4, 5]. Therefore, it is possible to mine drug use patterns with high frequency for standardized treatment and rational drug use by exploring data–driven methods from EMRs.

As the cost of healthcare is increasing more rapidly than the willingness and the ability to pay for it, and more and more data is being captured around healthcare processes in the form of EMRs, health insurance claims, and wearable data, data mining has become critical to the healthcare world [5]. Recently, a large number of studies have shown that secondary use of EMRs has enabled data-driven of clinical pathways discovery [6, 7], treatment pattern extraction [8–11], patient diseases and hospitalization prediction [12, 13], and patient stratification and subgroups identification [14]. Thus, it would be valuable to utilize massive EMRs to aid clinicians for supporting clinical decision making and improving the overall delivery of healthcare services [5, 15].

In order to effectively discovery knowledge from EMRs, patient similarity measure and clustering are essential [8–11]. Patient similarity measures how similar a pair of patients are according to their historical information recorded in the EMRs, such as demographic information, diagnostic information, laboratory examinations, and treatment information [16, 17]. In the existing researches, some patient similarity measure methods have been proposed, such as latent Dirichlet allocation [6], the doctor order content-based set similarity [8], typicalness index [10], doctor order sequence-based similarity [11], the vector space model [18], duration-aware pairwise trace alignment [19], and dynamic time warping [20]. And clustering is an important unsupervised learning technique of partitioning set of objects into multiple groups (called clusters) so that objects in the same cluster are more similar to each other than to those in other clusters [21]. Many clustering algorithms have been also proposed, such as K-means, spectral clustering, hierarchical clustering, density-peaks-based clustering, and affinity propagation (AP) clustering.

Furthermore, some similar researches have been proposed to extract typical treatment patterns from EMRs [8–11]. While, in this study, considering the distribution features for drug use, we formulate a new representation of patient treatment records and design a patient similarity measure method to extract typical drug use patterns, including typical drug use set, typical drug use day set, and the drug use distribution feature of each typical drug, and further evaluate and label typical drug use patterns based on demographic and diagnostic information. Experimental study on a real EMRs has also validated the effective of our proposed method to extract typical drug use patterns and recommend a standard treatment for new patients.

The remainder of the paper is organized as follows. Section 2 presents terms and definitions of patient treatment records and drug use distribution feature. Section 3 proposes the methods to extract, evaluate and label typical drug use patterns. Section 4 shows and interprets our experimental study. Finally, the conclusions and future work are provided in Sect. 5.

2 Terms and Definitions

A treatment record of a patient is a set of drugs prescribed by the clinicians. It can be represented as $Treatment = \left\{ d_{gt} | g \in \{1, \ldots, n\}, t \in \bigcup_{j=1}^{m} [T_{d_g}^{Start\ date}, T_{d_g}^{End\ date}]_j \right\}$ for each patient (e.g., N patients in the EMRs, $i \in \{1, \ldots, N\}$), where d_{gt} is the g-th drug used in the t-th day, $T_{d_g}^{Start\ date}$ and $T_{d_g}^{End\ date}$ are the start and end date of the d_g, m indicates that the d_g appears m times with different $[T_{d_g}^{Start\ date}, T_{d_g}^{End\ date}]$ (e.g., [2008/3/21, 2008/3/25] and [2008/3/23, 2008/3/28]) in EMRs, and n is the total number of drug use during the treatment process of the specific patient.

Considering the existence of many repeated $Start$ and End date for the same drug, the treatment can be redefined as $Treatment' = \{ d_{gt}(l(d_{gt})) | g \in \{1, \cdots, n\}, t \in T^O \}$, where the $l(d_{gt})$ indicates how many times the d_g is used for the specific patient per day, $T^O = unique(\bigcup_{j=1}^{m} [T_{d_g}^{Start\ date}, T_{d_g}^{End\ date}]_j)$. Additionally, we also define the lasting days T_{d_g} of the drug d_g, $T_{d_g}=length(T^O)$, where $T^O = [T_{d_g}^{t_1}, T_{d_g}^{t_2}, \ldots, T_{d_g}^{t_{T_{d_g}}}]$ denotes no repeated date.

In order to effectively mine typical drug use patterns from EMRs, we firstly define drug use distribution feature ($DUDF$) with a set of quintuple for each drug to study drug use regularities. For example, for the d_g, $DUDF_{d_g} = (u_{d_g}, \sigma_{d_g}^2, T_{d_g}, T_{d_g}^{min(Start\ date)}, T_{d_g}^{max(End\ date)})$, where u_{d_g} denotes the mean of drug use days, $u_{d_g} = \sum_{j \in \{1, 2, \cdots, T_{d_g}\}} T_{d_g}^{t_j} \bigg/ T_{d_g}$, $\sigma_{d_g}^2$ denotes the variance of drug use days, $\sigma_{d_g}^2 = E(t_j^d - u)^2 = \sum_{j \in \{1, 2, \cdots, T_{d_i}\}} (T_{d_g}^{t_j} - u_{d_g})^2 \bigg/ T_{d_g}$, and T_{d_g} denotes the lasting days of drug use, $T_{d_g} = length(T^O)$, while $T_{d_g}^{min(Start\ date)}$ and $T_{d_g}^{max(End\ date)}$ are the first and last day of drug use, i.e., $T_{d_g}^{t_1}$ and $T_{d_g}^{t_{T_{d_g}}}$, respectively. In particular, the definition $Treatment'$ of treatment record is a crucial step to obtain the $DUDF$, though the $l(d_{gt})$ is not used in the $DUDF$. Table 1 shows a simple example about how to obtain the $DUDF$ of drug (e.g., Calcium Gluconate) for three treatment records of patient hospital admissions (HADM) from Medical Information Mart for Intensive Care III (MIMIC-III).

Table 1. A simple example of $DUDF$ generation of Calcium Gluconate

HADM_ID	Drug	Drug use days										
		1	2	3	4	5	6	7	8	9	10	11
100028	Calcium Gluconate	×	√	√	×	×	×	×	×	×	×	×
100074	Calcium Gluconate	√	√	√	×	×	×	×	×	×	×	×
100104	Calcium Gluconate	√	√	√	√	√	√	√	√	√	√	√

Thus we can obtain $DUDF = \{2.5, 0.25, 2, 2, 3\}, \{2, 0.6667, 3, 1, 3\}$, and $\{6, 10, 11, 1, 11\}$ of Calcium Gluconate for three patients, respectively

3 Methods

In this section, we propose a research framework for extracting typical drug use patterns from patient treatment records. Specifically, the research framework consists of four steps in Fig. 1: (1) similarities measure between patient treatment records, including drug use distribution feature vector (*DUDFV*) representation and similarity computation between *DUDFV*s, (2) Patient treatment records clustering, (3) Typical drug use pattern extraction from each cluster, (4) Typical drug use pattern evaluation and labelling.

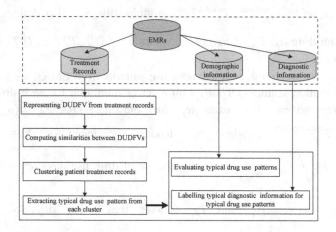

Fig. 1. A research framework for typical drug use pattern extraction

3.1 Similarity Measure Between Patient Treatment Records

Considering the definition *DUDF* of each drug, we design a novel similarity measure method to compute the similarity between pairwise patient treatment records.

(1) Representing *DUDFV* from patient treatment records
For a treatment record of a patient i, the $DUDF_{ig}$ of the g-th drug (i.e., d_g) can be obtained by

$$DUDF_{id_g} = \begin{cases} (u_{id_g}, \delta^2_{id_g}, T_{id_g}, T^{min(Start\ date)}_{id_g}, T^{max(End\ date)}_{id_g}) & if\ d_g \in T_i \\ (0, 0, 0, 0, 0) & Otherwise \end{cases} \quad (1)$$

Thus we can represent *DUDFV* of patient i based on the ordered combination of *DUDF*s for all drugs and define the *DUDFV* as

$$DUDFV_i = (U_i, \Sigma_i, T_i, T_i^{min(Start\ date)}, T_i^{max(End\ date)})_{1 \times 5M}, \tag{2}$$

where $U_i = (u_{i1}, \ldots, u_{iM})$, $\Sigma_i = (\delta_{i1}^2, \ldots, \delta_{iM}^2)$, $T_i = (T_{i1}, \ldots, T_{iM})$, $T_i^{min(Start\ date)} = (T_{i1}^{min(Start\ date)}, \ldots, T_{iM}^{min(Start\ date)})$, $T_i^{max(End\ date)} = (T_{i1}^{max(End\ date)}, \ldots, T_{iM}^{max(End\ date)})$, and M is the number of drugs for all patient treatment records.

(2) Computing similarities between DUDFVs

After representing *DUDFV* for each patient treatment record, we can form a matrix *DUDFVs* of $N * 5M$ dimension, where N is the number of patient treatment records. Firstly, in order to eliminate the dimensional effect between indicators, data standardization for each column (e.g., the h-column) of matrix *DUDFVs* is needed by

$$DUDFVs'(h) = \frac{DUDFVs(h) - min(DUDFVs(h))}{max(DUDFVs(h)) - min(DUDFVs(h))}, h = 1, 2 \ldots, 5M. \tag{3}$$

And the Euclidean distance between $DUDFV_i$ and $DUDFV_j$ can be measured by

$$dist(DUDFV_i, DUDFV_j) = (\sum_{h=1}^{5M} (DUDFVs_i'(h) - DUDFVs_j'(h))^2)^{1/2}, \tag{4}$$

where $DUDFVs_i'(h)$ denotes the value of i-th row and h-th column in the matrix *DUDFVs'*. Similarly, $DUDFVs_j'(h)$ is the value of j-th row and h-th column in the matrix *DUDFVs'* (i.e., patient j).

Thus, the similarity between $DUDFV_i$ and $DUDFV_j$ can be computed by

$$s(DUDFV_i, DUDFV_j) = 1/(1 + dist(DUDFV_i, DUDFV_j)). \tag{5}$$

And finally, we can obtain the similarity matrix S for all patient treatment records based on $s(T_i, T_j) = s(DUDFV_i, DUDFV_j)$.

3.2   Clustering Patient Treatment Records

In order to mine typical drug use patterns from large-scale treatment records, clustering algorithms is essential. After obtaining the patient similarity matrix S in Sect. 3.1, we can use clustering algorithms to divide all treatment records into several clusters. In this study, we adopt a popular exemplar-based clustering algorithm-AP clustering [22], which has some advantages than other clustering algorithms, such as not predefinition for the number of clusters, the real existence of the exemplars, and much lower error.

Furthermore, we use the input exemplar preferences (p_0) and the Sum of Similarities (SS) to select the proper number (K) of clusters. The former focuses on the robust of the clustering results than the direct predefinition of K, which depends on the similarity matrix S, the number N of patient treatment records, and adjustment parameter p^c (p coefficient) [22]. And it is defined as

$$p_0 = mean(S) - p^c \cdot N. \tag{6}$$

While the latter is a important criterion of exemplar-based clustering algorithm to evaluate the clustering performance, which depends on the similarity matrix S, the number N of patient treatment records, and the exemplar e for each cluster C [23]. And it is defined as

$$SS = \sum_{i=1}^{N} s(T_i, e_i). \tag{7}$$

A larger SS value shows a better clustering performance.

After clustering all the patient treatment records into K clusters (i.e., C_1, C_2, ..., C_K), we can obtain the *support* of each cluster as

$$Support(C_i) = \sum_{j=1}^{N} \delta(C(T_j), e_i) \Big/ N, \, i = 1, \ldots, K, \tag{8}$$

where $\delta(C(T_j), e_i) = 1$ if the cluster label of the treatment record T_j equals to the exemplar e_i of the cluster C_i, $\delta(C(T_j), e_i) = 0$, otherwise. A larger *support* value indicates that the cluster is more popular than other clusters.

3.3 Extracting Typical Drug Use Pattern from Each Cluster

In order to extract typical drug use patterns, we firstly define the core treatment records for each cluster as described in the literature [8, 11]. The aim of core treatment records is to mine some more representative patterns and avoid a large number of trivial patterns that appear in our results.

For the i-th cluster C_i, the core treatment records is defined by k-nearest neighbors (KNN) of its exemplar e_i as

$$Core_i = \{T_j | s_{T_j, e_i} \geq \tau_i\}, \, i = 1, \ldots, K \tag{9}$$

where τ_i denotes the similarity between the exemplar e_i and its k-th nearest neighbor, and T_j belongs to the i-th cluster C_i.

Then we can compute the *support* of each drug for the i-th cluster C_i based on the core treatment records $Core_i$ by

$$Support_i(drug_g) = \frac{\sum\limits_{T_j \subseteq Core_i} \lambda(drug_g, T_j)}{|Core_i|}, \, g = 1, \ldots, M, \tag{10}$$

where $\lambda(drug_g, T_j) = 1$ if $drug_g$ is used in the T_j, $\lambda(drug_g, T_j) = 0$ otherwise.

And we further define typical drug use set for each cluster according to the *support* of Eq. (10) as

$$TDrugSet_i = \{drug_g | Support_i(drug_g) > \delta_1\}, \tag{11}$$

where δ_1 is a threshold defined aforehand.

Finally, we can then compute the *support* of the p-th day for each typical drug (e.g., $drug_g$) from $TDrugSet_i$ by

$$Support_i(drug_{gp}) = \frac{\sum_{T_j \subseteq Core_i} \lambda(drug_g, T_{jp})}{|Core_i|}, p = 1, \ldots, P, \tag{12}$$

where $\lambda(drug_g, T_{jp}) = 1$ if $drug_g$ is used in the p-th day of the T_j, $\lambda(drug_g, T_j) = 0$ otherwise, T_{jp} represents the p-th day of the T_j, and P is the total drug use days.

And we further define effective drug use day set for each typical drug (e.g., $drug_g$) according to the *support* of Eq. (12) as

$$EDrugDaySet_{ig} = \{drug_{gp} | Support_i(drug_{gp}) > \delta_2\}, \tag{13}$$

where δ_2 is also a threshold defined aforehand.

Furthermore, according to the definition of the *DUDF* in Sect. 2 and the effective drug use day set of typical drug in Eq. (13), we can obtain the *DUDF* of each typical drug (e.g., $drug_g$) by

$$DUDF_{drug_g} = (u_{drug_g}, \delta^2_{drug_g}, d^T_{drug_g}, d^{min(Start\ date)}_{drug_g}, d^{max(End\ date)}_{drug_g}). \tag{14}$$

Thus we can extract the typical drug use pattern for each cluster including typical drug use set, typical drug use day set, and the *DUDF* of each typical drug.

3.4 Evaluating and Labelling Typical Drug Use Patterns

After obtaining the typical drug use patterns from patient treatment records, we further retrieve the demographic and diagnostic information of patients with core treatment records to analyze the characteristics of the extracted patterns. For demographic information, we select five kinds of representative characteristics including admission type (e.g., patient condition), age, gender, hospital mortality and average death days (within the hospital), total mortality and average death days (the sum of within and outside the hospital). For diagnostic information, we directly select the all ICD codes of core patients diagnosed by clinicians.

For the evaluation of typical drug use patterns, we can compute the *support* (i.e., occurrence probability) of the q-th category for each representative characteristic of demographic information by

$$Support_i(DeI_q) = \frac{\sum_{T_j \subseteq Core_i} \lambda(DeI_q, PDeI(T_j))}{|Core_i|}, q = 1, \ldots, Q, \tag{15}$$

where *DeI* denotes five representative characteristics of demographic information, such as admission type, age, gender, hospital mortality, and total mortality, $PDeI(T_j)$ denotes the given patient demographic information with T_j, $\lambda(DeI_q, PDeI(T_j)) = 1$ if DeI_q is occurred in the $PDeI(T_j)$, $\lambda(DeI_q, PDeI(T_j)) = 0$ otherwise, and Q is the number of categories for a specific *DeI*.

For the diagnostic information labelling of typical drug use patterns, we can then compute the *support* of each diagnostic ICD code by

$$Support_i(D\_ICD_h) = \frac{\sum\limits_{T_j \subseteq Core_i} \lambda(D\_ICD_h, PDiI(T_j))}{|Core_i|}, h = 1, \ldots, H, \qquad (16)$$

where D_ICD is all diagnostic ICD codes recorded in all core patient diagnostic information, $PDiI(T_j)$ denotes the given patient diagnostic information with T_j, $\lambda(D\_ICD_h, PDiI(T_j)) = 1$ if D_ICD_h is occurred in the $PDeI(T_j)$, $\lambda(D\_ICD_h, PDiI(T_j)) = 0$ otherwise, and H is the number of all diagnostic ICD codes.

And we further define the effective diagnostic ICD code set according to the *support* of Eq. (16) as

$$ED\_ICDSet_i = \{D\_ICD_h | Support_i(D\_ICD_h) > \delta_3\}, \qquad (17)$$

where δ_3 is also a threshold defined aforehand.

Thus we can adopt the occurrence probability of five representative characteristics extracted from demographic information to evaluate typical drug use patterns, and use the effective diagnostic ICD code set to label typical drug use patterns.

4　Experimental Study

In this section, a real-world EMR data is used to extract and evaluate typical drug use patterns. We first describe data collection, preprocessing and experimental setup, select the proper number of clusters, then extract typical drug use patterns from patient treatment records, and further evaluate and label the typical drug use patterns.

4.1　Data Collection, Preprocessing and Experimental Setup

In our experiment, we use a freely accessible critical care database: MIMIC-III [24], which has been widely conducted patient outcome improvement and health services research, such as mortality prediction in the ICU, early warning of clinical deterioration, patient cohort identification, and multiple diagnosis codes for ICU patients [15, 25], and then select sepsis patients from MIMIC-III as our experimental dataset. Sepsis is a potentially life-threatening complication of an infection, and has become a major and costly disease in the ICU [26, 27]. According to the disease severity, sepsis is classified into sepsis, severe sepsis and septic [28].

Figure 2 describes the detailed processes of data collection and preprocessing of sepsis patients from MIMIC-III database, including diagnostic information, treatment records and demographic information. Specifically, we firstly retrieve 5,172 sepsis patients by matching the sepsis disease with ICD_9 code, using the HADM_ID as patient coding, deleting the repeated patients, and selecting adult patients (aged 18 years or above) with discharge reports. Then, after a series of data preprocessing for patient treatment records, 5169 sepsis patients are obtained. Finally, we further extract the diagnostic information, treatment records and demographic information for these sepsis patients.

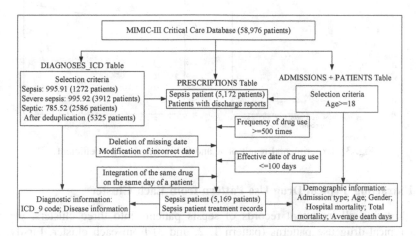

Fig. 2. Data collection and preprocessing of the experimental dataset

Furthermore, in order to effectively extract and evaluate typical drug use patterns, we also discuss the experimental setup in this section. Firstly, drug use (183 medications) and ICD_9 code (3,334 codes) of patients are recoded with numbers as "drug code" and "diagnosis code", respectively. Besides, admission type, age, and gender are classified into "elective", "urgent", and "emergency"; "below age 50 years" and "age 50 years or above"; and "female" and "male", respectively. Hospital mortality represents the death rate of patients within the hospital, while total mortality consists hospital mortality and the deaths identified by matching the patient to the social security master death index. Average death days is the lasting survival time after treatment, including the death within the hospital and the death within and outside the hospital. Next, the selection of cluster number (K) is also discussed. Finally, the k is set 200 in Eq. (9) to define a dense core for each cluster, the δ_1 is set 0.5 in Eq. (11) to define typical drug use set, the δ_2 is set 0.2 in Eq. (13) to define typical drug use day set, and the δ_3 is set 0.3, 0.4, and 0.5 in Eq. (17) to define the effective diagnostic ICD_9 code set.

4.2 Selection of Cluster Number (K)

After obtaining the similarity matrix S for 5169 sepsis patient treatment records according to the method described in Sect. 3.1, the AP cluster algorithm with sparse capability (apclusterSparse.m[1]) is applied to divided all the treatment records into several clusters. Figure 3 shows the experimental result of K and SS by adjusting the different p^c (p coefficient) in Eqs. (6) and (7) to select a proper K. From Fig. 3, it has a more robust clustering result with a large SS when p^c ranges 0.0025 to 0.004. Therefore, we finally select 3 clusters ($p^c = 0.003$) to extract typical drug use patterns.

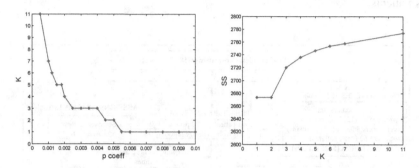

Fig. 3. Experimental result of K and SS in terms of p coefficient

4.3 Extracting Typical Drug Use Pattern from Each Cluster

After clustering all treatment records of sepsis patients into three clusters, we can extract typical drug use patterns (pattern 1, 2, and 3) from each cluster. Firstly, we extract the core treatment records ($k = 200$) from three clusters based on Eq. (9), and calculate the total treatment days for each patient, as shown in Fig. 4(1), where the horizontal denotes the core treatment records recoded with numbers as "patient code"[2], and the vertical coordinate is the total drug use days for each "patient code". Then we can also calculate the support of each drug (183 drugs) and select the typical drug use set ($\delta_1 = 0.5$) based on Eqs. (9) and (10) from each pattern, as shown in Fig. 4(2–4).

In Fig. 4(1), there exists a larger difference among three patterns for total drug use days, where the pattern 1 is the shortest ranging from 1 to 4 days, the pattern 3 is medium from 2 to 5 days, while the pattern 2 is the longest from 5 to 15 days. Additionally, we find that the overall trend of average drug use days for three patterns is rising with the decrease of similarity. In Fig. 4(2–4), obviously, the pattern 2 uses the maximum number of typical drugs with a higher support, the pattern 3 is medium, while the pattern 1 is the minimum with a lower support. Moreover, the pattern 2 and 3 use a lot of the same typical drugs with the drug type of main and additive, such as 3^3 (Potassium Chloride), 9 (Vancomycin), 18 (Magnesium Sulfate), 12 (Sodium Chloride 0.9% Flush), 13 (Acetaminophen), and 22 (Norepinephrine), while the drug type of base is different except 4 (Iso-Osmotic Dextrose). The same typical drugs used in three patterns are Iso-Osmotic Dextrose and 14 (Heparin). Finally, we further select three

---

[1] http://www.psi.toronto.edu/index.php?q=affinity%20propagation.

[2] The first code is the exemplar of a cluster, the last is the smallest similarity with the exemplar.

[3] The 3 denotes that the drug code 3 in Fig. 4(3) is Potassium Chloride.

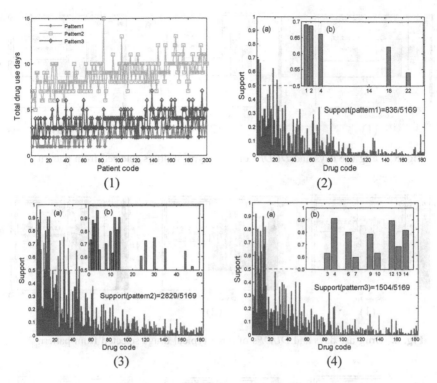

Fig. 4. The core treatment records and typical drug use set extracted from 3 clusters.

typical drugs to analyze their effective drug use days and *DUDF*s from each pattern, respectively.

Figures 5, 6 and 7 show the effective drug use days and *DUDF*s of three typical drugs. Obviously, the typical drugs used in the pattern 2 have the longest effective drug use days ($\delta_2 = 0.2$) with a higher support, the pattern 3 have the medium effective drug use days, and the pattern 1 have the shortest effective drug use days with a lower support. Additionally, the *DUDF*s of three typical drugs for each pattern are the similar. While for the specific typical drugs, the effective drug use days and *DUDF*s are very different for three patterns.

4.4 Evaluating and Labeling Typical Drug Use Patterns

Section 4.3 has shown that our proposed method can extract typical drug use patterns from core treatment records. In this section, we further evaluate and label typical drug use patterns with demographic and diagnostic information, considering that the same typical drug use pattern of sepsis patients should have the similar admission information, disease severity and treatment outcome. Therefore, for three typical drug use patterns, Table 2 shows a detailed description of demographic information of core patients ($k = 200$), Fig. 8 shows the distribution of diagnostic information of core patients.

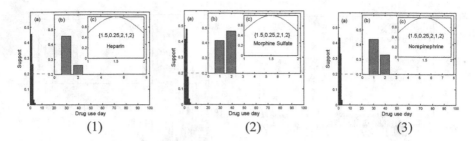

Fig. 5. Drug use days and *DUDF*s of three typical drugs selected from pattern 1

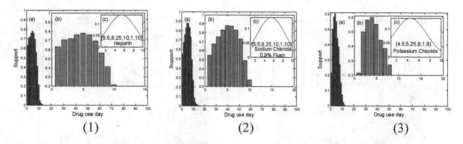

Fig. 6. Drug use days and *DUDF*s of three typical drugs selected from pattern 2

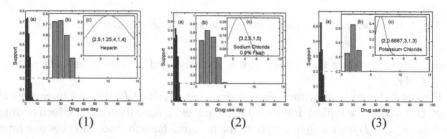

Fig. 7. Drug use days and *DUDF*s of three typical drugs selected from pattern 3

In Table 2, in general, we find that for all the core patients extracted from three clusters, the admission information is the similar, such as the largest support of admission type with emergency (over 98.5%) and age 50 years or above (over 85%), and almost no difference in gender, but the treatment outcomes are very different. Specifically, for patents in pattern 1, the hospital and total mortality almost reach 100% but with the shortest average death day, possibly indicating that a less typical drugs in Fig. 4(2) and effective drug use days in Fig. 5. For patients in pattern 2, the hospital mortality is the lowest with the longest average death days of about one week within the hospital, possibly because of a lot of typical drugs used in Fig. 4(2) and the longer effective drug use days in Fig. 6. While the total mortality (mainly outside the hospital) is rising rapidly when patients discharge from hospital with average death days of about

Table 2. The detailed description of demographic information for three patterns

Pattern	Admission type			Age		Gender		Mortality*	Average death day
	A	B	C	<50	≥ 50	Female	Male		
P1	0	1.5%	**98.5%**	9.5%	**90.5%**	47.5%	52.5%	**90.5%/ 94%**	**1.35/6.52**
P2	1%	0.5%	**98.5%**	11.5%	**88.5%**	55%	45%	**14%/ 61.5%**	**7.53/347**
P3	0	1%	**99%**	15%	**85%**	43.5%	56.5%	**26%/47%**	**2.1/147**

P1, P2, and P3 denote Pattern1, Pattern2 and Pattern3, respectively. A, B, and C denote Elective, Urgent, and Emergency, respectively. Mortality denotes the Hospital mortality and Total mortality.

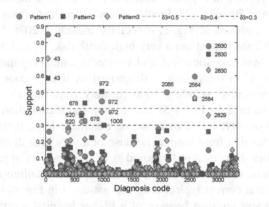

Fig. 8. The distribution of diagnostic information for three patterns

one year. And for patients in pattern 3, the mortality within hospital and outside the hospital is the similar with average death days of two days and half a year, respectively.

In Fig. 8, all the effective diagnostic ICD_9 code sets under different patterns and δ_3 values (0.3, 0.4, and 0.5) are marked with diagnostic code, such as 43 (septicemia NOS with ICD_9 code 3.89), 620 (hypertension NOS), 678 (congestive heart failure NOS), 972 (acute kidney failure NOS), 1006 (urinary tract infection NOS), 2086 (acute respiratory failure), 2584 (septic shock), 2829 (sepsis), and 2830 (severe sepsis), where NOS refers to the unspecified essential diseases. First of all, we find that more than 3000 diagnostic ICD codes co-exist with sepsis disease, and some common complications can be identified, such as septicemia, acute respiratory failure, acute kidney failure, hypertension, congestive heart failure, and urinary tract infection. Secondly, the effective diagnostic ICD_9 code sets can't be identified when δ_3 values is less than 0.3, because the distributions of diagnostic ICD_9 code for three patterns are the same and trivial, thus the minimum δ_3 value should be set to 0.3. And then, the main diagnostic ICD_9 codes for three patterns are the same when δ_3 value is set to 0.3, except acute respiratory failure in pattern 1, urinary tract infection NOS in pattern 2, and sepsis in pattern 3. Next, when δ_3 value is set to 0.4, the main diagnostic ICD codes for pattern 1 and pattern 2 are also the same, except acute respiratory failure and congestive heart failure NOS, and the disease severity of sepsis patients in pattern 1 and pattern 2 are more serious than that in pattern 3. Finally, when δ_3 value is set to 0.5, the different of diagnostic ICD codes is obvious, and the disease severity of sepsis patients in pattern 1 is more serious than that in pattern 2. Therefore, in general, considering the number and support of the effective diagnostic ICD code sets and different δ_3 value, we deem the disease severity of sepsis patients is the most serious in pattern 1, the more serious in pattern 2, and the serious in pattern 3.

4.5   Clinical Decision Support for New Patients

In medical practice, the extracted typical drug use patterns from treatment records can assist clinicians to make a standardized treatment regimen within and outside the hospital. Specifically, for a new patient admitted to ICU, the admission type is emergency, and age is over 50 years old. Firstly, when the patient is diagnosed as ICD_9 code set of pattern 1 shown in Fig. 8, especially including 2086 (acute respiratory failure), we deem that the patient has a very high death risk, typical drug use pattern 1 in Fig. 4(2) has a little therapeutic effect, and a better treatment regimen may need to be explored. Secondly, when the patient is diagnosed as ICD_9 code set of pattern 2 shown in Fig. 8, especially including 1006 (urinary tract infection), we deem that typical drug use pattern 2 in Fig. 4(3) has a very significant therapeutic effect for reducing hospital mortality, and a more active treatment (e.g., pattern 2) should be continued after discharging from hospital because of a very high death risk outside the hospital. Finally, when the patient is diagnosed as ICD_9 code set of pattern 3 shown in Fig. 8, without serious complications, like acute respiratory failure and urinary tract infection, we deem that current typical drug use pattern 3 in Fig. 4(4) may be not the most effective treatment regimen because of a higher hospital mortality, thus a new treatment regimen (e.g., pattern 2) should be also explored by clinicians. Besides, the treatment should also be continued after discharging from hospital because of a higher death risk outside the hospital.

5   Conclusion

In this paper, we design a novel research framework to mine and extract typical drug use patterns from patient treatment records. Specifically, considering the drug distribution regularities, we first define *DUDF* with a set of quintuple for each drug and further represent patient treatment record with *DUDFV*. Then we propose a similarity measure method to compute the similarity between patient treatment records. Next we use AP clustering algorithm to cluster all patient treatment records, extract a typical drug use pattern from each cluster, and evaluate and label the extracted typical drug use patterns with demographic and diagnostic information. Finally, we adopt our proposed approach on a real EMR data of sepsis patients collected from MIMIC-III database. Experimental results indicate that our approach can effectively extract typical drug use patterns from patient treatment records, and assist clinicians to design a better treatment regimen based on their demographic and diagnostic information admitted to hospital.

However, there also exist several issues in our study. Due to the lack of real label of patient treatment records, the importance of each feature in *DUDF* and the comparison with other similarity methods are not considered. In our future work, we will further take into account the label of treatment records under the guidance of clinicians to optimize the weight of each feature in *DUDF* by genetic algorithm, compare our method with state-of-the-art techniques, and validate the extracted results with some professionals in the medical area to increase the value of our research.

Acknowledgements. This work was supported in part by the Natural Science Foundation of China [Grant Numbers 71771034, 71421001] and China Postdoctoral Science Foundation [Grant Number 2017M620054].

References

1. Desalegn, A.A.: Assessment of drug use pattern using WHO prescribing indicators at Hawassa University teaching and referral hospital, south Ethiopia: a cross-sectional study. BMC Health Serv. Res. **13**(1), 170 (2013)
2. WHO: The Rational Use of Drugs. Report of a conference of experts, Nairobi, 25–29 November 1985. World Health Organization, Geneva (1987)
3. Khan, S.U., Zomaya, A.Y., Abbas, A.: Handbook of Large-Scale Distributed Computing in Smart Healthcare. Springer, New York (2017). https://doi.org/10.1007/978-3-319-58280-1
4. Jensen, P.B., Jensen, L.J., Brunak, S.: Mining electronic health records: towards better research applications and clinical care. Nat. Rev. Genet. **13**(6), 395–405 (2012)
5. Yadav, P., Steinbach, M., Kumar, V., Simon, G.: Mining electronic health records (EHR): a survey. ACM Comput. Surv. **50**(6), 1–40 (2018)
6. Huang, Z., Dong, W., Bath, P., Ji, L., Duan, H.: On mining latent treatment patterns from electronic medical record. Data Min. Knowl. Discov. **29**(4), 914–949 (2015)
7. Perer, A., Wang, F., Hu, J.: Mining and exploring care pathways from electronic medical records with visual analytics. J. Biomed. Inf. **56**, 369–378 (2015)
8. Sun, L., Liu, C., Guo, C., Xie, Y., Xiong, H.: Data-driven automatic treatment regimen development and recommendation. In: Proceedings of the 22rd ACM SIGKDD International Conference on Knowledge Discovery and Data Mining, pp. 1865–1874. ACM (2016)
9. Yang, S., Dong, X., Sun, L., Zhou, Y., Farneth, R.A., Xiong, H.: A data-driven process recommender framework. In: Proceedings of the 23rd ACM SIGKDD International Conference on Knowledge Discovery and Data Mining, pp. 2111–2120. ACM (2017)
10. Hirano, S., Tsumoto, S.: Mining typical order sequences from EHR for building clinical pathways. In: Peng, W.C., et al. (eds.) PAKDD 2014. LNCS, vol. 8643, pp. 39–49. Springer, Cham (2014). https://doi.org/10.1007/978-3-319-13186-3_5
11. Chen, J., Sun, L., Guo, C., Wei, W., Xie, Y.: A data-driven framework of typical treatment process extraction and evaluation. J. Biomed. Inf. **83**, 178–195 (2018)
12. Liu, C., Wang, F., Hu, J., Xiong, H.: Temporal phenotyping from longitudinal electronic health records: a graph based framework. In: Proceedings of the 21st ACM SIGKDD International Conference on Knowledge Discovery and Data Mining, pp. 705–714. ACM (2015)
13. Riccardo, M., Li, L., Kidd, B.A., Dudley, J.T.: Deep patient: an unsupervised representation to predict the future of patients from the electronic health records. Sci. Rep. **6**, 26094 (2016)
14. Li, L., et al.: Identification of type 2 diabetes subgroups through topological analysis of patient similarity. Sci. Transl. Med. **7**(311), 311ra174–311ra174 (2015)

15. Data, M.I.T.C.: Secondary Analysis of Electronic Health Records. Springer, New York (2016). https://doi.org/10.1007/978-3-319-43742-2
16. Sun, J., Wang, F., Hu, J., Ebadollahi, S.: Supervised patient similarity measure of heterogeneous patient records. ACM SIGKDD Explor. Newsl. 14(1), 16–24 (2012)
17. Wang, F., Sun, J., Ebadollahi, S.: Integrating distance metrics learned from multiple experts and its application in patient similarity assessment. In: Proceedings of the 2011 SIAM International Conference on Data Mining. Society for Industrial and Applied Mathematics, pp. 59–70 (2011)
18. Garcelon, N., et al.: Finding patients using similarity measures in a rare diseases-oriented clinical data warehouse: Dr. warehouse and the needle in the needle stack. J. Biomed. I. 73, 51–61(2017)
19. Yang, S., et al.: Duration-aware alignment of process traces. In: Perner, P. (ed.) ICDM 2016. LNCS (LNAI), vol. 9728, pp. 379–393. Springer, Cham (2016). https://doi.org/10.1007/978-3-319-41561-1_28
20. Forestier, G., Lalys, F., Riffaud, L., Trelhu, B., Jannin, P.: Classification of surgical processes using dynamic time warping. J. Biomed. Inf. 45, 255–264 (2012)
21. Han, J., Kamber, M., Pei, J.: Data Mining: Concepts and Techniques, 3rd edn. Morgan Kaufmann Publishers Inc., Burlington (2011)
22. Frey, B.J., Dueck, D.: Clustering by passing messages between data points. Science 315 (5814), 972–976 (2007)
23. Sun, L., Guo, C., Liu, C., Xiong, H.: Fast affinity propagation clustering based on incomplete similarity matrix. Knowl. Inf. Syst. 51(3), 1–23 (2016)
24. Johnson, A.E.W., Pollard, T.J., Shen, L., Lehman, L.W.H., Feng, M., Ghassemi, M., et al.: MIMIC-III, a freely accessible critical care database. Sci. Data 3, 160035 (2016)
25. Wang, S., Li, X., Chang, X., Yao, L., Sheng, Q.Z., Long, G.: Learning multiple diagnosis codes for ICU patients with local disease correlation mining. ACM Trans. Knowl. Discov. Data 11(3), 31 (2017)
26. Johnson, A., Stone, D.J., Celi, L.A., Pollard, T.J.: The MIMIC code repository: enabling reproducibility in critical care research. J. Am. Med. Inf. Assoc. 25(1), 32–39 (2017)
27. Martin, G.S.: Sepsis, severe sepsis and septic shock: changes in incidence, pathogens and outcomes. Expert Rev. Anti-Infect. Ther. 10(6), 701–706 (2012)
28. Singer, M., et al.: The third international consensus definitions for sepsis and septic shock (Sepsis-3). J. Am. Med. Assoc. 315(8), 775–787 (2016)

Aiding First Incident Responders Using a Decision Support System Based on Live Drone Feeds

Jerico Moeyersons[✉][iD], Pieter-Jan Maenhaut[iD], Filip De Turck[iD], and Bruno Volckaert[iD]

Department of Information Technology, IDLab Ghent University - imec,
Technologiepark-Zwijnaarde 15, 9052 Ghent, Belgium
Jerico.Moeyersons@UGent.be

Abstract. In case of a dangerous incident, such as a fire, a collision or an earthquake, a lot of contextual data is available for the first incident responders when handling this incident. Based on this data, a commander on scene or dispatchers need to make split-second decisions to get a good overview on the situation and to avoid further injuries or risks. Therefore, we propose a decision support system that can aid incident responders on scene in prioritizing the rescue efforts that need to be addressed. The system collects relevant data from a custom designed drone by detecting objects such as firefighters, fires, victims, fuel tanks, etc. The drone autonomously observes the incident area, and based on the detected information it proposes a prioritized based action list on e.g. urgency or danger to incident responders.

This paper presents the architecture of the framework and a prototype implementation and evaluation of a decision support system, responsible for digesting and prioritizing the large amount of contextual data captured at an incident site. The evaluation of the decision support system shows that the proposed solution works accurately in supporting incident responders in providing a sorted overview of the actions needed in real-time, with an average response time of 334 ms on a less powerful device and 263 ms on a powerful device equipped with a GPU.

Keywords: Disaster management · Recommendation systems
Decision analysis and decision support systems

1 Introduction

In 2013, a train carrying chemical products derailed in Wetteren, Belgium [23]. Because of this derailment, chemical products contained in the wagons exploded, causing an inferno and the release of toxic substances in the air. The fire department initially used water to extinguish the fire, resulting in a chemical reaction with the released contents and the spread of toxic water through the sewers of

© Springer Nature Singapore Pte Ltd. 2018
J. Chen et al. (Eds.): KSS 2018, CCIS 949, pp. 87–100, 2018.
https://doi.org/10.1007/978-981-13-3149-7_7

a nearby city. This disaster caused one death and several injuries, all of them intoxicated by the toxic gases rising from the sewer system. If the fire department had known that the wagons contained chemical products, they would have used foam instead of water to douse the flames, resulting in no spreading of toxic water through the sewers and most likely fewer victims.

To prevent such scenarios in the future, we propose a solution in which a drone supports first incident responders. The drone can fly autonomously to the incident area and circles around the incident area multiple times to scan and analyze it by detecting different relevant objects. This results in a more rapid and better overview of the incident as a whole and, when connected to an automated decision support system, also aids in making split second decisions regarding evacuation, on-scene toxic materials (through object recognition), etc. The main goal of the proposed system is to save lives. The automated decision support is responsible for capturing the huge amount of context and historical data available during such an event and analyzing it, in order to provide incident responders on scene with a better overview of the situation, and to offer dangerous situation warnings and an intervention action list ordered according to urgency. The actual decisions however are made by incident responders and the presented system therefore serves as a decision support system.

In this paper, we primarily focus on the design of a novel decision support process, which takes as input different types of object detections provided by airborne drones and other contextual information (e.g. weather information, database information about gas pipelines). Afterwards we evaluate this decision support process to check whether it can provide real-time decisions in case of an incident. The remainder of this paper is organized as follows. In Sect. 2 related work is discussed, followed by an overview of the overall architecture in Sect. 3. In Sect. 4 the decision support system is explained in-depth and subsequently evaluated in Sect. 5. Finally, conclusions and avenues for future work are provided in Sect. 6.

2 Related Work

Several publications focus on how to improve the Emergency Response Management (ERM) with IT solutions. Banjeree et al. [5] for example describe web and mobile applications that are developed for disaster management. They conclude that many applications and frameworks guarantee better critical communication, situational awareness, resource overview and management, and improved decision support. Because every solution focuses on one of previous improvements, the authors propose a unified framework to combine the different solutions. Their disaster management focuses on alerting civilians in case of a greater event. Our solution on the other hand focuses on alerting incident responders based on detailed real-time context-information (stemming a.o. from a custom autonomous drone), and considering the role of the incident responder, e.g. a firefighter may need to be in closer vicinity of a live fire than a paramedic.

Concerning chemical disasters, Dotson et al. [10] propose a framework for detection and decision making. The decision making is based on existing rule databases in combination with learning from previous incidents. In our paper, we use a deep neural network for detection and decision making instead of existing rule databases. The approach utilized by Dotson et al. for prioritizing detections and decisions could be useful in eventual further work.

Chen et al. [8] discuss different approaches on how ERM can be improved. They propose a non-IT framework focused on better communication and better situational overview. The target of their framework is that other researchers use it to improve ERM, by understanding the overarching requirements for coordination design and implementation. The conclusions from this work summarize different aspects that are needed when building a whole ERM system and has been marked for future work.

The department of Homeland Security built a middleware platform, called Unified Incident Command and Decision Support (UICDS), described by Morentz et al. [18]. UICDS collects data from both governmental and commercial incident management technologies, combines these and creates a role-based, situational awareness information sharing platform. UICDS primarily focuses on alerting civilians in case of specific events to achieve better risk prevention, protection, response and recovery. The proposed deployment methods of UICDS can be used to deploy our decision support system and the UICDS middleware platform gives a good overview of new components that can be added.

Other approaches for supporting incident responders are based on the use of crowd sourcing information, described in Laskey [16] and Abu-Elkheir et al. [4]. The use of pictures, videos and status updates on social media can offer insights to first incident responders during an emergency event. Both solutions propose a way to implement a crowd sourcing platform, but the main bottleneck with this approach is to verify which data is applicable and which data must be discarded, such as outdated information or fake news. Crowd sourcing information can provide an additional, highly interesting source of information for our decision support system, and has been marked for further research.

In Kotsiantis [15] several classification methods based on machine learning are compared and explained. Table 1 gives a simplified overview of the different features of a selection of learning techniques. It is clear that a Neural Network (NN) and a Support Vector Machine (SVM) have better accuracy, but slower training times compared with a Rule Learner (RL) and a Decision Tree (DT). An important difference between SVM methods and the others is that SVM methods are binary, and thus in the case of multi-class problem one must reduce the problem to a set of multiple binary classification problems. Important features for the described decision support system in this paper are accuracy in general, the speed of a decision, the attempts of incremental learning (e.g. the system can learn from different datasets with the intent to get better each time) and the possibility to handle a multi-class problem (a situation needs to be classified into a clear, warning or danger situation). The decision support system proposed in this paper extends these approaches in order to design, prototype and evaluate a decision-making framework supporting incident responders in making split-second decisions in case of an emergency.

It is clear research into decision support systems for incident response is ongoing. On the first hand however some solutions are very incident specific, e.g. for chemical disasters while on the other hand other solutions are focusing on alerting nearby civilians in case of a large incident. Our decision support system focuses on alerting and aiding incident responders on site during a generic event with a potential to plug in incident-specific and civilian alerting solutions.

Table 1. Overview of different learning algorithms. (**** stars represent the best and * star represent the worst performance)

	Decision trees [17,19,24]	Neural networks [17]	Support vector machines [7,9]	Rule learners [11–13]
Accuracy in general	**	***	****	**
Speed of learning	***	*	*	**
Speed of decision	****	****	****	****
Tolerance to missing values	***	*	**	**
Tolerance to highly interdependent attributes	**	***	***	**
Danger of overfitting	**	*	**	**
Attempts of incremental learning	**	***	**	*
Solving multi-class problems	****	****	**	****

3   Architecture Overview

Figure 1 provides a general overview of the system. A custom-built drone, visualized in Fig. 2, is used, equipped with both a FLIR BOSON thermal camera [1] and two IMX274 4K cameras [2] used for stereoscopic imaging. The drone autonomously detects all relevant objects in the area, such as the emergency workers, the location of the fire, dangerous objects (e.g. barrels containing explosive fluids) etc. The image processing happens on the drone itself, using a Nvidia Jetson TX2 GPU board, and object detection and localization is implemented using the methods presented in our previous work [21,22].

The contextual information captured by the drone is forwarded to the back-end for further processing, together with the original video feed and additional drone telemetry (location, heading, altitude, battery level, etc.). The back-end analyses this data in real time, using a decision support system. Alongside current feeds, historical data from previous interventions is considered. Furthermore, these data feeds are stored in the back-end for future use (e.g. to learn from

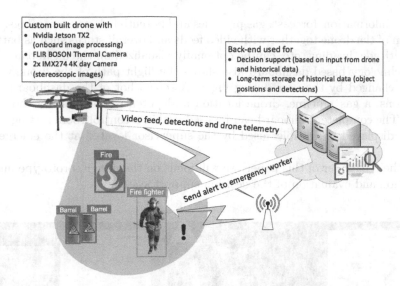

Fig. 1. General overview of the autonomous drone-based decision support system for incident response.

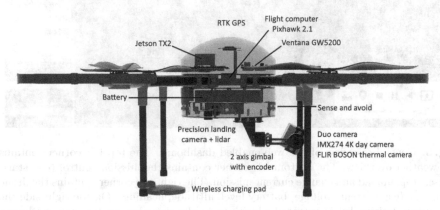

Fig. 2. General overview of the custom build drone.

mistakes made while dealing with a crisis situation and potentially introduce new rules avoiding such situations in the future). Decision making processing on the drone itself is not recommended as object recognition and communication already demand a lot of processing power/energy on the drone causing an extra stress on the battery which will results in shorter flight time. When the decision support system detects a dangerous situation, e.g. when a fire is located near explosive materials, an alert is generated and forwarded to all emergency workers located in the danger zone.

The back-end also provides a user-friendly dashboard in which all identified objects are visualized on a map, together with other relevant information such as weather information/forecast, gas pipelines and potential alerts or warnings. The location of the drone together with video feeds and overall status such as battery level, altitude, heading, etc. can be optionally visualized on the dashboard. The web-technology based dashboard also shows the flight path of the drone which can be changed by the user during flight. A screenshot of the dashboard with detections, a gas pipeline, drone location and potential decisions is shown in Fig. 3. The complete dashboard can be used within a control room setting, or it can be displayed on a tablet used by the supervisor located at the emergency site.

In the remainder of this paper, we will focus on the design, prototype implementation and evaluation of the decision support system.

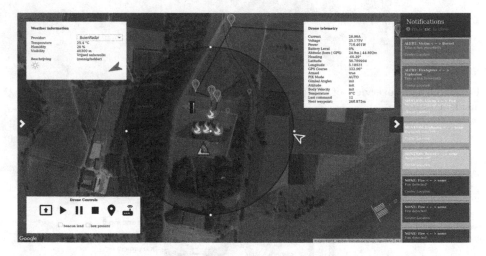

Fig. 3. General overview of the web-enabled dashboard. The top left corner contains the weather overview. The bottom left corner contains the mission control (e.g. start, pause, stop, upload and change current mission. The right top corner contains the drone telemetry (e.g. current, voltage, battery level, altitude, heading). On the right side the prioritized decision list is provided where a red color describes urgent actions (danger situations), a yellow color describes a warning situation and a gray color describes an informative message. The current drone location is marked with a white arrow marker and the blue lines show the flight path of the drone. In the Point-of-Interest circle (part of the drone flight path) the current detections are marked and labeled with a self-explainable marker. (Color figure online)

4   Decision Support - TensorFlow

The decision support system is implemented using TensorFlow [3], an open-source, state-of-the-art machine learning framework created by Google. Tensor-Flow contains a wide range of functionality, and can therefore be used in a wide area of applications, but is mainly designed for creating a Deep Neural Network (DNN). The reason to opt for a DNN instead of the other proposed learning methods in Sect. 2 is related to the specific use case in this paper. The use case under investigation requires an accurate and fast approach of a multi-class problem that can be distributed and re-trained. Eventual training speed is less important because training can happen between different interventions on a computer or server with stronger hardware. If we compare these requirements with the features of the different learning methods in Table 1, a DNN was picked as a suitable candidate. The used DNN for the decision support system consists of three layers, consisting of 100, 200 and 100 neurons respectively, whereby the whole network is dense. The used activation function is the Rectified Linear Unit (ReLu) activation function [20], the most successful in terms of results and widely used activation function. It gives an output x when the input (x) is positive and 0 otherwise, resulting in fewer activated neurons in the DNN which means less computational overhead and thus faster results for the incident responders. The used classifier function is the Softmax cross entropy function [6]. It squashes a K-dimensional vector of arbitrary real values to a K-dimensional vector of real values in the range (0, 1) that add up to 1 (the probabilities). This means that the commander and dispatcher can see how certain the decision support system is about its result and eventually can take that into account when making a final decision. A result with a higher probability is more likely to be correct than one with a lower probability.

The input for the DNN are two detected objects and the distance between them, all detected and calculated by the drone discussed in Sect. 3. Note that other, non-distance related inputs are possible e.g. to generate an alert when a victim is detected by the drone. For easier calculations within the neural network, these input values, and especially the two detections are converted into a label with the Labelencoder from the Python sklearn package. Afterwards, these categorical integer labels are transformed with a one-hot, also called an one-of-K scheme. When applied to a scenario with four possible detections, this results in 9 input neurons, namely four for the first detection, four for the second detection and one for the distance between them. Figure 4 provides a general overview of the functioning of the DNN.

The output (prediction) of the DNN is the classification of the input into one of the three possible classes, namely a danger, warning or normal situation. These predictions, together with their probability is shown in an ordered fashion to the user (i.e. the commander at scene or in the dispatch center), to support them in gaining control of the crisis situation. The user can dismiss specific warnings or mark dangerous situations which are not flagged by the system.

After training the DNN, the model can be exported to be used as a *servable* within the TensorFlow Serving API [14]. The TensorFlow serving API allows to serve different machine learning models and manages their lifetimes, so clients can access versioned machine learning models via a high-performance, reference-counted lookup-table. This allows to train multiple different models and request a prediction from a specific model, managed by the TensorFlow server API. In case that there is no reliable communication channel between the back-end and the incident responders on scene, the model can be deployed on a local device in e.g. the intervention truck.

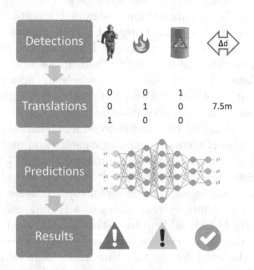

Fig. 4. General functioning of the Deep Neural Network: detections and distance between detections are translated and fed into a neural network which leads to classification of events into dangerous, potentially dangerous or safe occurrences.

5   Evaluation Results

Now that we have the architecture and the decision support system explained, an evaluation is needed to check whether the system can give decisions accurately and in real-time.

5.1   Datasets

The datasets used for the evaluation are built based on previously defined rules, created and generated under the supervision of a fire department expert active in the 3DSafeGuard-VL project. The expert was an ICT responsible for the fire department and cooperated in the design of different incident response protocols. In these rules, two detected objects and a specific distance between them result in different possible outputs, an alert in case of a danger situation, a warning in case of a potentially unsafe situation or nothing when no threats are detected. A sample of select entries in the dataset is provided in Table 2. The datasets are built only for evaluation purposes and the whole system is not limited to the detections specified below. As explained in Sect. 3 an autonomous incident-scanning drone provides the decision support system with real-time object detection capability.

Three different datasets are created to evaluate the change in accuracy and in response time when more rules are trained. Dataset 1 trains the decision support system on 190 different rules, based on 20 possible detected objects. Dataset 2 trains the decision support system on 435 rules, based on 30 possible detected objects and Dataset 3 trains the system on 990 rules, based on 45 possible detected objects. The total number of rules per dataset is based on the number of possible detected objects, namely $\sum_{i=1}^{x-1} i$ where x is the number of possible detected objects. An overview of the datasets is showed in Table 3.

Accuracy is important to the incident responders because they base their decisions in part on the predictions made by the decision support system. A lower accuracy means that there is a higher possibility of a wrong prediction. The three datasets have a different amount of rules to study their impact on the accuracy. With more rules incident responders can get more contextual information about the incident because more objects are scanned and analyzed. The response time on the other hand is also of great importance to evaluate the decision support system is capable of providing decisions in real-time or near real-time as a function of the number of rules in the system.

Table 2. Sample of dataset entries

Detection 1	Detection 2	Distance	Result
Firefighter	Explosion	10	Alert
Firefighter	Fire	5	None
Victim	Fire	10	Warning
Victim	Explosion	5	Alert
Fire	Explosion	2	Alert

Table 3. Overview of the different datasets. (# pos. detections shows how many different objects can be identified such as a firefighter, fire, smoke, etc., # training entries shows on how many entries of the dataset training is done and # evaluation items shows on how many entries the accuracy of the model is determined.)

Name	# Rules	# pos. detections	# training items	# evaluation items
Dataset 1	190	20	7600	1520
Dataset 2	435	30	18560	3712
Dataset 3	990	45	40760	8152

5.2   Results

Several experiments were performed using the three different datasets summarized in Table 3. First, the training phase is evaluated with a varied range of training steps, going from 100 to 40000 in steps of 100 in function of the obtained accuracy. This training is done on a Nvidia GTX 980 Ti GPU with CUDA 9.0 and TensorFlow 1.8 installed and the results are visualized in Fig. 5. The accuracy of the DNN is at the highest levels starting from 4500 training steps. More training steps do not guarantee better accuracy in general. An important conclusion from Fig. 5 is that a model with less rules has better accuracy than a model with more rules. This behavior can be explained because there are less possibilities within the NN itself, causing a lower fault rate and thus a higher accuracy.

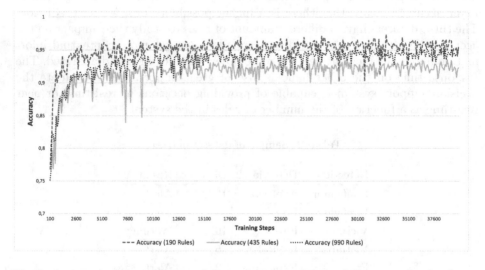

Fig. 5. Accuracy of the trained model as a function of the training steps for the three datasets consisting of 190, 435 and 990 rules.

After training three different DNN-models from the datasets, these models are exported to another (less powerful) laptop. This laptop contains 8 GB of RAM, an Intel-i5 processor and lacks a GPU. On this laptop, the models are exposed to predict 36 decisions, each based on 2 detections and the distance between them. This will indicate that a model can be exported to a local device in e.g. an intervention truck when communication with the back-end is not reliable or available. When the local device is connected to the drone, it is thus possible to have decision support on site. As shown in Fig. 6, the response time is equally for the three different models. The total time the three models, consisting of 190, 435 and 990 rules, needed for making 36 predictions are 421 ms, 408 ms and 432 ms respectively. It is proven that a model works concurrent and can handle multiple request at the same time. A possible explanation for the outliers in response time is that the request was stalled when the model was predicting another request.

Figure 7 shows the response time on 36 predictions for the three models on a device with a GTX 980Ti equipped. Compared to the device that lacks a GPU,

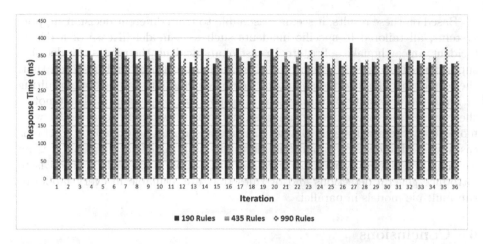

Fig. 6. Required time (in ms) for calculating a prediction on a CPU based on 2 detections and the distance between them, evaluated on the three models consisting of 190, 435 and 990 rules. The request are sent after each other and the total time for predicting these 36 request are 421 ms, 408 ms and 432 ms respectively.

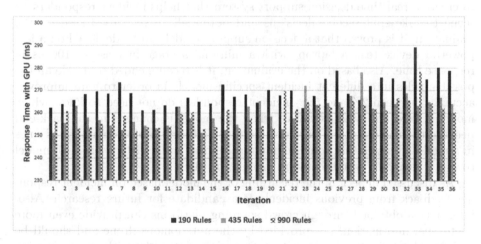

Fig. 7. Required time (in ms) for calculating a prediction on a GPU based on 2 detections and the distance between them, evaluated on the three models consisting of 190, 435 and 990 rules. The request are sent after each other and the total time for predicting these 36 request are 315 ms, 307 ms and 306 ms respectively.

the average response time is 23.59% faster for each individual response as for the 36 responses together. It is also remarkable, when using a system with a GPU, that the response time for 190 rules is higher in most cases than the response time for the models with 435 and 990 rules. When the model has more rules, more Nvidia-CUDA cores will be used in parallel, which result in lower response times. This is highly dependent on the specification of the used GPU.

Based on these results, it seems reasonable to train different decision support systems, all tailored to specific incidents such as a model in case of a diving incident, an earthquake, a fire, etc. In case of an incident, the request for decision support needs to specify which type of incident is occurring and based on that, the correct model is selected and more accurate predictions can be done resulting in improved reliability of the system. The incident responders will not notice the change from one model to another, and when switching to a model with more rules, there will be no change in response time. The drawback to this is cases where the decision support needs to aid with mixed incidents such as an earthquake followed by multiple fires and explosions. A possible solution is to introduce general detections and or rules that are available in every model or run multiple models in parallel.

6 Conclusions

This paper proposed the use of a decision support system for the classification of potential dangers based on data retrieved from a drone autonomously surveying an incident zone. Prototyped with TensorFlow, it was shown that it is possible to create a real-time decision support system that helps incident responders on scene to take split second decisions and potentially save lives. Based on the evaluation, it is proven that a decision support model can be deployed on a less powerful device (e.g. a laptop) with a minimal average increase of 100 ms in response time. Also based on the evaluation, it is recommended to divide all the possible rules into different incident-specific models in order to gain improved accuracy between 2% and 5%. Incident responders do not need to know which model is used and will not notice when the model is changed in the back-end because the response time does not increase as function of the number of rules in a model. Remark that some common rules need to be included in each model to avoid having to switch between models too often.

As discussed in Sect. 3, the relearning of the model based on incident responder feedback from previous incidents is a candidate for future research. Also, information obtained through crowd sourcing platforms can provide even more contextual information, not provided by the autonomous drone and should be tied into the decision support system. One of the difficulties with crowd sourcing remains however autonomous selection of relevant data. Finally, a system will be introduced which will allow for more warning handling prioritization on top of the classification between danger, warning or a clear situation.

Acknowledgement. The authors would like to thank all partners namely, Xtendit Solutions, Fire.BE, DroneMatrix, CityMesh, Seris Security, SAIT and KU Leuven - Eavise, in the 3DSafeGuard-VL project and the Agency for Innovation by Science and Technology (Vlaio) for funding and support.

References

1. FLIR Boson. http://www.flir.it/cores/boson/

2. Sony Semiconductor Solutions Corporation - IMX274QC. https://www.sony-semicon.co.jp/products_en/new_pro/october_2016/imx274lqc_e.html
3. Abadi, M., et al.: TensorFlow: a system for large-scale machine learning. In: 12th USENIX Symposium on Operating Systems Design and Implementation (OSDI 2016), pp. 265–284 (2016). https://doi.org/10.1038/nn.3331
4. Abu-Elkheir, M., Hassanein, H.S., Oteafy, S.M.: Enhancing emergency response systems through leveraging crowdsensing and heterogeneous data. In: 2016 International Wireless Communications and Mobile Computing Conference, IWCMC 2016, pp. 188–193 (2016). https://doi.org/10.1109/IWCMC.2016.7577055
5. Banerjee, A., Basak, J., Roy, S., Bandyopadhyay, S.: Towards a collaborative disaster management service framework using mobile and web applications. Int. J. Inf. Syst. Crisis Response Manag. 8(1), 65–84 (2016). https://doi.org/10.4018/IJISCRAM.2016010104
6. Bishop, C.M.: Pattern Recognition and Machine Learning. Springer, New York (2006). https://doi.org/10.1117/1.2819119
7. Burges, C.J.: A tutorial on support vector machines for pattern recognition. Data Min. Knowl. Discov. 2(2), 121–167 (1998). https://doi.org/10.1023/A:1009715923555
8. Chen, R., Sharman, R., Rao, H.R., Upadhyaya, S.J.: Coordination in emergency response management. Commun. ACM 51(5), 66–73 (2008). https://doi.org/10.1145/1342327.1342340
9. Cristianini, N., Shawe-Taylor, J.: An Introduction to Support Vector Machines: And Other Kernel-based Learning Methods. Cambridge University Press, New York (2000)
10. Dotson, G.S., Hudson, N.L., Maier, A.: A decision support framework for characterizing and managing dermal exposures to chemicals during Emergency Management and Operations. J. Emerg. Manag. 13(4), 359–380 (2015). https://doi.org/10.5055/jem.2015.0248
11. Fürnkranz, J.: Pruning algorithms for rule learning. Mach. Learn. 27(2), 139–172 (1997). https://doi.org/10.1023/A:1007329424533
12. Fürnkranz, J.: Separate-and-conquer rule learning. Artif. Intell. Rev. 13(1), 3–54 (1999). https://doi.org/10.1023/A:1006524209794
13. Fürnkranz, J.: Round robin rule learning. In: Proceedings of the Eighteenth International Conference on Machine Learning, ICML 2001, pp. 146–153. Morgan Kaufmann Publishers Inc., San Francisco (2001). http://dl.acm.org/citation.cfm?id=645530.655685
14. Google Research: Tensorflow Serving (2017). tensorflow.github.io/serving/
15. Kotsiantis, S.B.: Supervised machine learning: a review of classification techniques. Informatica 31, 249–268 (2007). https://doi.org/10.1115/1.1559160
16. Laskey, K.B.: Crowdsourced decision support for emergency responders 4444(703) (2013)
17. Lim, T.S., Loh, W.Y., Shih, Y.S.: A comparison of prediction accuracy, complexity, and training time of thirty-three old and new classification algorithms. Mach. Learn. 40(3), 203–228 (2000). https://doi.org/10.1023/A:1007608224229
18. Morentz, J.W., Doyle, C., Skelly, L., Adam, N.: Unified incident command and decision support (UICDS): a department of homeland security initiative in information sharing. In: 2009 IEEE Conference on Technologies for Homeland Security, HST 2009, pp. 182–187 (2009). https://doi.org/10.1109/THS.2009.5168032
19. Murthy, S.K.: Automatic construction of decision trees from data: a multidisciplinary survey. Data Min. Knowl. Discov. 2(4), 345–389 (1998). https://doi.org/10.1023/A:1009744630224

20. Nair, V., Hinton, G.E.: Rectified linear units improve restricted boltzmann machines. In: Proceedings of the 27th International Conference on International Conference on Machine Learning, ICML 2010, Omnipress, USA, pp. 807–814 (2010). http://dl.acm.org/citation.cfm?id=3104322.3104425
21. Tijtgat, N., Ranst, W.V., Volckaert, B., Goedemé, T., De Turck, F.: Embedded real-time object detection for a UAV warning system. In: 2017 IEEE International Conference on Computer Vision Workshops (ICCVW), pp. 2110–2118 (2017). https://doi.org/10.1109/ICCVW.2017.247
22. Tijtgat, N., Volckaert, B., De Turck, F.: Real-time hazard symbol detection and localization using UAV imagery. In: 2017 IEEE 86th Vehicular Technology Conference (VTC-Fall), pp. 1–5 (2017). https://doi.org/10.1109/VTCFall.2017.8288259
23. Torfs, M.: 1 dead, 33 injured in Wetteren train derailment, May 2013. http://deredactie.be/cm/vrtnieuws.english/News/1.1620582
24. Yildiz, O.T., Dikmen, O.: Parallel univariate decision trees. Pattern Recogn. Lett. 28(7), 825–832 (2007). https://doi.org/10.1016/j.patrec.2006.11.009

The Use of Event-Based Modeling and System-Dynamics Modeling in Accident and Disaster Investigation

Xiangting Chen[1(✉)] and Xiao Liu[2]

[1] Carnegie Mellon University, Pittsburgh, PA 15213, USA
xiangtic@andrew.cmu.edu
[2] Jinan University, Guangzhou 510632, Guangdong, China
lxchdd@jnu.edu.cn

Abstract. The classical accident and disaster investigation process is centered around modeling the sequence of events that lead to failure. In the Big Data Era, such event-based analyses benefit from the diverse sources of accident-related data and powerful data analytics techniques. However, it is argued that a system-dynamics perspective is also crucial in understanding the dynamic evolution of complex accident and disaster systems. In this paper, the Integrated Event-Based Modeling and System-Dynamics Modeling (EBSD) Framework for Accident Investigation using Big Data Analytics is proposed, and a case study of the 2015 Tianjin Port Fire and Explosion is presented to demonstrate its potential application. The EBSD Framework provides assistance to investigators in extracting the key factors in the disaster and accident system through modeling and simulation, and thus have further implication in disaster and accident management.

Keywords: Disaster and accident management · Accident investigation
System dynamics modeling · Tianjin port fire and explosion

1 Introduction

Disaster and accident management had been an important focus in our society, mainly due to the fact that disasters bring substantial economic losses to government and corporations and cause psychological trauma to individuals. Take the fire and explosion that occurred at Rui Hai International Logistics Company, which was located in Tianjin Port, China, on August 12th, 2015, as an example: the casualty report was 165 deaths, 8 missing, and 798 non-fatal injuries, and the estimated direct finial loss was 6.8 billion yuan (about 1 billion U.S. dollar) [1]. Given the profound negative impacts of such disasters and accidents, it is important to conduct accident investigation, whose goal is to determine the sequence of events that lead to failures, to extract the main causes of accidents, and most importantly, to give recommendations to stakeholders such as corporate managers and policy makers, so that they can build robust systems and develop preventive regulations.

The purpose of accident investigation is to identify events and causal factors leading to the accident [2]. Many efforts and thinking have been put to refining and

© Springer Nature Singapore Pte Ltd. 2018
J. Chen et al. (Eds.): KSS 2018, CCIS 949, pp. 101–114, 2018.
https://doi.org/10.1007/978-981-13-3149-7_8

optimizing the accident investigation process, resulting in an impressive number of well-established methods in the literature, each of which has its own strengths and contributes to the whole process of understanding an accident and deducing its root cause [3, 4]. Domino Theory, proposed by Heinrich in the 1940s, was one of the aerialist accident causation models [5]. It describes an accident as a chain of discrete events that occur in a particular order. Following the advent of Domino Theory, many sequential accident models and methodologies were invented, including Failure Modes and Effects Analysis (FMEA), Fault Tree Analysis (FTA), Event Tree Analysis, and Cause-Consequence Analysis, all of which belong to the class of event-based accident models. These models work well in relatively simple systems that involves physical components failure and/or human errors [3].

However, by only examining events in a linear and sequential fashion, the traditional approach to accident investigation ignores the effect of feedback and complex interactions between system variables and failed to acknowledge the existence of causal chains that could produce complex behaviors through processes such as feedback [6, 7]. Therefore, accident investigators should also adopt a system theoretic approach [8], which utilize the tool of system dynamics, a field emerged in the 1950s that now has applications in a wild range of areas such as population study and supply chain management. Such an approach requires investigators to move away from looking at isolated events and to start to think about accidents in terms of a complex interacting system that includes feedback loops, time delays, and non-linear relationships. It is important to note that system dynamics modeling is not independent of the classical approaches in accident analyses; instead, it serves as a continuum of event-based task analyses that allows us to simulate how different factors in an accident system interact with each other and evolve dynamically over time. It follows that a new paradigm for disaster and accident analyses that combines these two approaches is desired.

The classical accident investigation process can be summarized into three main phases: (i) collecting evidence and facts; (ii) analyzing evidence and facts and drawing conclusions; (iii) developing judgment and need and writing the report [9]. In phase two, which is also the most important one, many investigators focus on identifying different types of errors (e.g., human error, task error, material/equipment error, etc.) and constructing event-based accident models. The classical investigation process, or disaster management (DM) as a whole, benefits greatly from the advent of the Big Data Era, which made available various forms of data from many different sources. In the meantime, Big Data Analytics (BDA) techniques also provide people with a more powerful way of conducting accident analyses, in turn helping people understand present accident and predict future ones [10, 11]. For instance, texting mining methods were applied to extract key causal relations from maritime accident investigation reports collected from the Marine Accident Investigation Branch [12].

This paper is organized as the following: In Sect. 2, the Integrated Event-Based Modeling and System-Dynamics Modeling (EBSD) Framework for Accident Investigation using Big Data Analytics is proposed and explained. In Sect. 3, a case study of the Tianjin Port Explosion using the proposed methodology is presented in detail. Limitation and future work is described in the last section.

2 Research Design

With the advent of the Big Data Era, new research paradigms based on big data analytics emerged in many different fields of study, including the area of disaster and accident investigation, which is one of the important fields of safety science [10, 11]. In their paper, Huang et al. proposed a big data-based accident investigation paradigm, which seeks to apply data analytic methodologies to accident-related data gathered from various sources. Although the attempt to incorporate big-data analytics to the investigation process was novel, their paradigm relies on the classical approach in accident analyses. As it is presented in the previous section that a system-dynamics modeling approach is also needed, the Integrated Event-Based Modeling and System-Dynamics Modeling (EBSD) Framework for Accident Investigation using Big Data Analytics is proposed (Fig. 1).

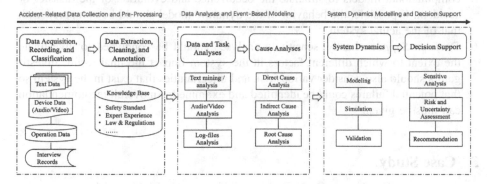

Fig. 1. EBSD framework for accident investigation using big data analytics

There are six major steps in the EBSD Framework. The first and second steps focus on gathering data from numerous sources and pre-processing them for later analyses. The next two steps use traditional accident analyses tools and methods while having the benefit of utilizing the power of big data analytics. The fifth step extends the classical cause analyses to system dynamics modeling and simulation. Lastly, through conducting simulation and sensitivity analyses, recommendations could be made, and conclusions could be drawn. A detailed description of the different stages is presented below:

1. *Data acquisition, recording, and classification.* In this stage, data from various sources are acquired, recorded, and extracted; they should be further classified into several categories (e.g., text, audio, and video, etc.) based on their form.
2. *Data extraction, cleaning, and annotation.* The goal of this step is to put the data in either a structured or unstructured form that is suitable for further analyses. Methods include various data pre-processing techniques.
3. *Data and Task Analyses.* In this step, depending on the content type of the data, text analytics, which refers to techniques that extracts information from textual data,

audio analytics, which refers to techniques that analyze and extract information from unstructured audio data, and video analytics, which refers to techniques that monitor, analyze, and extract information from video streams, should be applied [11]. Then, with the pooled information, traditional event-based task analysis methods, such as event tree analysis and fault tree analysis, should be used [2].

4. *Cause Analyses*. This step follows naturally from the previous one, and it is often the last steps in traditional event-based accident analysis models. It is advised that Causal Loop Diagrams (CLDs) are used to represent the causal relationships and feedback structures that exist in the accident case; they are vital in helping investigators to think about the case systematically [13].

5. *System Dynamics Modeling*. CLDs can then be transformed into Stock and Flow Diagrams (SFDs), which is another tool used by system scientists to study the behavior of dynamic systems [13]. Using data gathered in previous steps and making assumptions and simplifications when necessary, investigators can then use computational models to simulate the occurrence and evolution of the disaster or accident. If possible, the behavior of the dynamic model should be compared with that of the actual system.

6. *Decision Support*. Lastly, sensitivity analyses should be conducted to investigate the extent to which different factors in the system impact the final result. Investigators should also consider various risk and uncertainties that exist in the system so that potential failures could be identified and examined. An overall recommendation can thus be given.

3   Case Study

In this section, a case study of the 2015 Tianjin Port fire and explosion accident using the EBSD Framework is presented. Previous investigations on the case have utilized many methodologies, such as a process analysis approach that is based on event sequence-barrier failed [14]. In addition, Big Data analytic methodologies were also applied to analyze unstructured data such as surveillance video footage [11]. Note that this case study only demonstrates a simplified use of the full framework since not all aspects were addressed. Nevertheless, the example should be able to convince the readers that the proposed framework is able to provide them with a comprehensive, cohesive, and clear map of the steps in conducting a successful accident and disaster investigation.

3.1   Acquisition, Recording, and Classification; Extraction, Cleaning, and Annotation

At 10:51 p.m. on August 12th, 2015, a fire broke out at the dangerous goods storage area of Rui Hai International Logistics Company, which lead to two major explosions that resulted in more than 150 deaths and a 6.8-billion-yuan financial loss. The final accident investigation report is given by the State Council of the People's Republic of China [1], which makes it an authoritative and reliable source of information. Thus, in this study, for the sake of simplicity, the first two steps are collapsed into a summary of the key findings given in the written report. According to the report, the series of events that lead to the onset of fire were:

1. Nitrocellulose became dry because of the mishandling of seal packages containing moisturizers;
2. Under high temperature in intermodal containers, dry Nitrocellulose started to decompose quickly, emitting a large amount of heat;
3. Because the containers are not effective in dissipating heat, dry Nitrocellulose continued to decompose and emit heat, to the point that the temperature reached its ignition point.

And the series of events described below lead to the two major explosions

1. A portion of dry Nitrocellulose in a particular intermodal container was on fire.
2. The small fire then ignited the remaining Nitrocellulose in that container, which resulted in a sharp increase in temperature and pressure.
3. The high temperature and pressure cause the container to explode, thus scattering the burning Nitrocellulose to the surrounding containers, which all stores dangerous chemicals, including refined naphthalene, sodium sulfide, trichlorosilane.
4. Eventually, the fire reached the container storing Ammonium Nitrate, which is also a violate chemical that would explode at a temperate above 400 °C. According to later simulations, after burning for about 30 min, Nitrocellulose would reach over 1000 °C, far exceeding the temperature that will cause Ammonium Nitrate to explode.
5. At 11:34:06, the first explosion occurred. The shock waves of the explosion and spreading fire affected the containers that stored more explosive chemicals located further away, leading to the second explosion.

3.2 Data and Task Analyses

There are two major Human Errors that lead to the disaster

1. Staffs at Rui Hai were extremely careless when loading and unloading dangerous goods, which caused the packages containing Nitrocellulose to break during the process.
2. Immediately after the accident had occurred, the staffs at Rui Hai didn't alert nearby companies to evacuate, thereby increasing the number of casualties.

Other errors are categorized into five distinct categories and are present below:

1. Task
   a. Disregarding safety regulations, Rui Hai International put chemicals with different properties and different levels of volatility together when storing.
   b. Disregarding safety regulations, Rui Hai International put intermodal containers too close to each other and even stacked 4 to 5 containers on top of each other.
   c. Disregarding government policy, Rui Hai International didn't do a safety evaluation of the location at which they store dangerous chemicals and didn't report to the government about storing such amount of dangerous chemicals.

d. Rui Hai International Logistics company violated related regulations regarding the maximum amount of dangerous chemicals a company can store. On the day of the accident, Rui Hai has about 1232.8 tons of Potassium Nitrate, 484 tons of Sodium Nitrate, and 680.5 Sodium Cyanide, which are 53.7 times, 19.4 times, and 42.5 times more than the legal amount, respectively.

2. Material
   a. There was a total of 72 types of dangerous chemical storing at Rui Hai International, including Nitrocellulose, which is the chemical that causes the initial fire, refined naphthalene, sodium sulfide, trichlorosilane, and Ammonium Nitrate. Many of the chemicals are volatile.
   b. There is no cooling system within each intermodal container.
3. Environment
   a. The temperature on the day of the accident was about 36 °C, under which the air temperature within intermodal containers can reach 65 °C, which exceeds the temperature for dry Nitrocellulose to decompose violently and emitting heat energy.
4. Individual Factors
   a. Some staffs in charge of loading and managing dangerous goods didn't receive any safety education and didn't obtain related certificate. They have very limited knowledge about how to prevent and handle accidents.
5. Management, Organizational, and Cultural
   a. Rui Hai International didn't establish any safety regulations regarding the process of loading and unloading. Thus, there is also no professionals overlooking the process of loading and unloading dangerous goods.
   b. Several government officials took bribes and gave Rui Hai International permission to store dangerous goods bypassing official procedures when they know that the company was incapable of handling a large amount of dangerous chemical.
   c. Government officials didn't ensure that Rui Hai International maintain their storage level within a legal limit.
   d. Government officials didn't ensure that Rui Hai International was storing dangerous good properly (e.g., not putting containers too close to each other and not mix-storing different chemicals).

The Fault Tree is presented on the following page (Fig. 2).

3.3 Cause Analyses Using Causal Loop Diagram

Figure 3 presents the Causal Loop Diagram created for this disaster. It is separated into two major parts, one that concerns the explosion, and one that focuses on the fire, and they are connected through the two variables "Probability of Fire" and "Probability of Explosion".

The explosion section of the CLD contains one balancing loop, and its behavior can be described as the following: the more profit-maximizing the company is, the greater the number of explosives it will store in the storage area; the greater the amount of goods that the company stores, the smaller the gap between the amount that it is storing

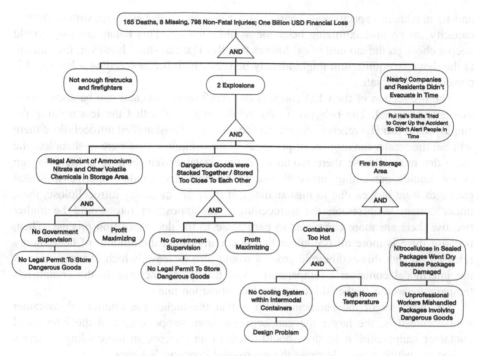

Fig. 2. Fault tree diagram

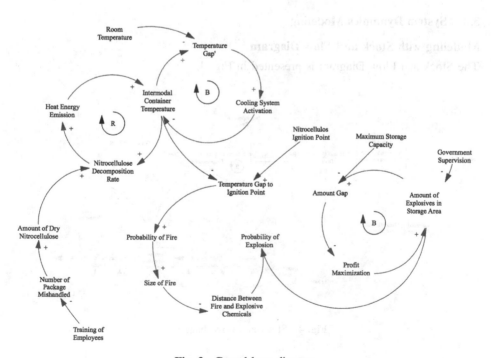

Fig. 3. Causal loop diagram

and its maximum capacity will be; and as the company reaches its maximum storage capacity, its profit-maximizing behavior would decrease. This balancing loop would keep a check on the amount of explosives that Rui Hai can store; however, the amount at the dynamic equilibrium might already be pretty high (i.e., exceeding a level that is considered to be safe).

The fire section of the CLD contains one reinforcing loop and one balancing loop that are connected. The behavior of the reinforcing loop is that the less training the employees at Rui Hai receive, the greater the number of mishandled nitrocellulose there will be; the greater the number of packages of nitrocellulose that were mishandled, the more dry nitrocellulose there would be if the container temperature is high enough (since liquids preventing nitrocellulose to become dry would vaporize if sealed packages were broken due to mishandling); if there are more dry nitrocellulose, then, under certain temperature, the nitrocellulose decomposition rate would be higher because there are more chemicals to participate in the decomposition process. Then, because there is more nitrocellulose that is decomposing, the amount of heat energy released through this exothermic process would also increase, which in turn increases the intermodal container's temperature. And finally, an increase in the container's temperature would further drive up the decomposition rate.

The behavior of the balancing loop is that the higher the intermodal container temperature it is, the larger the gap between room temperature and the intermodal container temperature it is; this should result in an increase in the cooling system's activation, which would decrease the intermodal container's temperature.

3.4   System Dynamics Modeling

Modeling with Stock and Flow Diagram

The Stock and Flow Diagram is presented in Fig. 4.

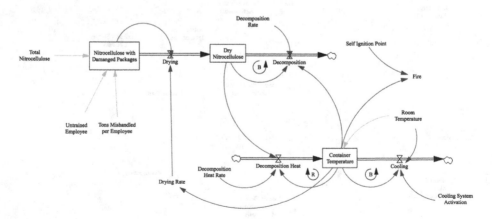

Fig. 4. Stock and flow diagram

Simulation using VensimPLE[1]

Experiment One: Effect of Cooling System

As it is mentioned above, the containers storing dangerous chemicals do not have a cooling system. In other words, it has a high danger of creating a positive reinforcing loop in which the container's inner temperature would rise exponentially. Using the relevant information that I've gathered regarding the case and making some informed assumptions (explained in the equations for each variable), I established a baseline model. The simulation results are presented below (Fig. 5).

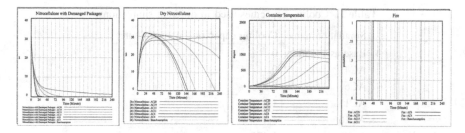

Fig. 5. Simulation output: baseline model for cooling system experiment

As we can see from the output graph, since there is no outflow for Container Temperature, it rises exponentially from 0 min to about 144 min after simulation. Note that the reason why the temperature stopped to rise further is that the rise in temperature comes from the heat energy released from the decomposition of dry nitrocellulose; as we can see from the graph for the amount of Dry Nitrocellulose, at about 144 min, its amount decreased to zero, putting a check to further rise in temperature. We can also see that at about 36 min into the simulation, the fire variable equals to one, which is an indication for the onset of fire in the containers.

The shape of output graph for the amount of Nitrocellulose with Damaged Packages shows goal-seeking behavior: the slope was first extremely negative but then gradually increased as the amount of nitrocellulose with damaged packages approaches zero. The behavior of the amount of Dry Nitrocellulose can be explained as the following: the amount first increased exponentially because the inflow, drying of the nitrocellulose with damaged packages, dominates the outflow, the decomposition.

However, after all the nitrocellulose with damaged packages turned into dry nitrocellulose, the inflow decreased to zero and the outflow thus dominates. Coupled with rising container temperature, the amount of dry nitrocellulose decreased sharply.

The following plots (Fig. 6) show the effect of having a cooling system with different capacities. The cooling capacities for different simulation are presented in Table 1.

As it is shown on Fig. 6, only for AC20 (blue line), where the cooling capacity was set to 1.5 °C/min, the container temperature did not reach the self-ignition point of the

---

[1] All parameters and equations used in the simulation are presented in the Appendix.

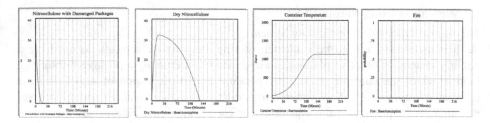

Fig. 6. Simulation output: cooling system experiment (Color figure online)

Table 1. Cooling capacities used in different simulations

Simulation name	Cooling capacity (degree/minute)
Base assumption	0
AC4	0.3
AC6	0.5
AC11	1.0
AC15	1.4
Ac19	1.475
AC20	1.5

dry nitrocellulose. In addition, we also observe that the greater the cooling capacity, the longer it takes for the container to reach the ignition point.

Experiment 2: Effect of Training of Employees

Through further examining the Fault Tree of the accident case, another important source of error that possibly led to the initial fire in the storage area is the mishandling of packages storing nitrocellulose. As it is shown in the following baseline simulation (Fig. 7), if there were no packages of nitrocellulose that were damaged, then nothing would have had happened. This is because, even though the intermodal container's inner temperature is high, the nitrocellulose would not become dry and would not catch fire because it is stored in sealed (undamaged) packages.

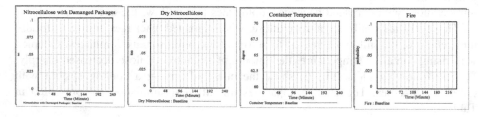

Fig. 7. Simulation output: untrained employee baseline model

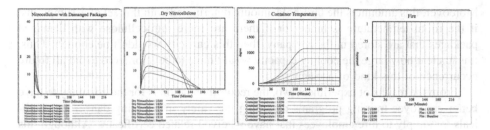

Fig. 8. Simulation output: untrained employee baseline model

Table 2. Percentage of untrained employees used in different simulations

Simulation name	Percentage of employees that received insufficient training	Number of untrained employees
Baseline	0%	0
UE10	10%	7
UE20	20%	14
UE30	30%	21
UE40	40%	28
UE50	50%	36
UE60	60%	43

The following plots show the effect of having different proportions of employees that received insufficient training, with the assumption that each untrained employee can mishandle and break the packages of one ton of nitrocellulose (Fig. 8). The different percentages for different simulation are presented in Table 2 (presented above).

As it is shown on the above simulation plots, we can observe that, under this set of assumptions, if the 20% or more of the employees at Rui Hai received insufficient training and thus mishandled packages storing nitrocellulose, a fire would have had occurred. In addition, we can also notice that in general, the more untrained employees there are, the faster it takes for the container temperature to reach the ignition point of Nitrocellulose; this is because an increase in the number of untrained employees would increase the amount of nitrocellulose with damaged packages, and thus potentially increases the amount of dry nitrocellulose, whose decomposition process could drive up the container temperature.

4   Decision Support

Recommendation
Based on the first experiment, we see that installing a cooling system inside the intermodal containers is one possible way of avoiding future accidents of the same kind. However, this solution is naïve since the threshold cooling capacity for which that could stop the onset of fire is about 1.5 °C/min, which is nearly impossible to reach.

Moreover, this amount is the threshold amount determined with this set of assumptions; if the room temperature is higher or if there are more nitrocellulose with damaged packages that are stored in the company, then this capacity wouldn't be enough. Thus, it is unclear how large the cooling capacity would have to be in order to avoid all future occurrence of this kind of accidents.

From the results of experiment two, the recommendation is for the company to have better employee training programs regarding the safety in handling dangerous chemicals, or simply to hire employees that have received related certificates. Although this approach might seem costly or time-consuming, it is able to produce a long-lasting and stable change. If there are no employees with little training that would break the packages containing nitrocellulose, then there would be no materials that would become dry and catch fire, regardless of the various environmental conditions that are often uncontrollable.

5   Conclusion

In this paper, the Integrated Event-Based Modeling and System-Dynamics Modeling (EBSD) Framework for Accident Investigation using Big Data Analytics was proposed, which combines modeling methodologies from both an event-based and a system-dynamics perspective; it also enables the use of big data analytics to extract intelligence from various sources. The goal of this framework is to assist investigators in seeing how different components of the accident or disaster system interact dynamically to produce the results, and in extracting the most important influencing factors.

The case study on the 2015 Tianjin Port Fire and Explosion Accident presents the real-world application of the EBSD Framework. Using both event-based modeling and system dynamics modeling techniques to analyze the case, we were able to extract two major human errors, which are (1) failure to install coolers in the intermodal containers that store volatile chemicals, and (2) mishandling of seal packages that contain flammable materials. However, it is only with the aid of the computation-based system dynamics modeling that we were able to compare different possible remedies and narrowed down the recommendation to be the more feasible solution of providing more rigorous employee training programs to reduce package-mishandling cases. This example demonstrates perfectly the value of a system dynamics perspective in accident investigation: by considering complex feedback structures and non-linear relationship, which is not possible with conventional event-based analysis, we gain insights into the interacting components that could lead to disaster and thus learn valuable lessons for future disaster management.

One limitation of the case study is the lack of the use of Big Data Analytics, which has a growing role in accident and disaster management. Since many information related to a disaster case might be recorded in text formats such as interview reports and system logs, text mining, which refers to the process of extracting information of interest from text, is an important Big Data Analytic technique. Text mining involves processes including Information Retrieval (IR), Information Extraction (IE), and Natural Language Processing (NLP). Further work is needed to better incorporate text mining methods into accident and disaster investigation process.

Acknowledgement. We want to acknowledge Dr. Cleotilde Gonzalez in introducing the concepts and research methodologies of task analysis and system dynamics. We also want to thank Erin McCormick for helping us constructing the CLDs and SFDs. The equations and simulations wouldn't have worked without them.

Appendix: Table of Equations and Parameters

Name	Parameters/Equations	Note
Total Nitrocellulose	32.97 (tons) [1]	
Self-Ignition Point	174 (degrees) [1]	
Room Temperature	35 (degrees) [1]	
Cooling System Activation	0 (degree/minute) [1]	There were no cooling systems within containers [1]
Untrained Employees	72 (people)	Assumption: All employees at Rui-Hai (72 in total) were untrained.
Tons Mishandled per Employee	1 (ton/people)	Assumption: No related data.
Decomposition Rate	0.0005 (ton/minute/degree)	Assumption: No related data.
Drying Rate	With Lookup: Container Temperature Lookup: ([(0,0) - (2000,1)], (0,0), (50,0.0083333), (175,1), (2000,1))	At 50 degrees, it takes 2 hours for Nitrocellulose to dry [1]; thus, the drying rate at 50 degrees is 1/120. Assumption: After the temperature has reached Nitrocellulose's self-ignition point, they would dry immediately.
Decomposition Heat Rate	0.00097	After burning for 30 min, nitrocellulose reached 1000 degree [1]. Assumption: Assuming that temperature rises exponentially, calculated the decomposition heat rate using a starting temperature of 175 degrees (self ignition point), an end temperature of 1000 degrees, and a burning time of 30 minute.
Nitrocellulose with Damaged Packages (tons)	Initial Value = MAX (0, MIN (Total Nitrocellulose, Untrained Employee * Tons Mishandled per Employee)) Equation = - Drying	
Drying (tons/minute)	IF THEN ELSE (Nitrocellulose with Damaged Packages <= 0, 0, MIN (Nitrocellulose with Damaged Packages, Drying Rate * Nitrocellulose with Damaged Packages))	
Dry Nitrocellulose (tons)	Initial Value = 0 Equation = Drying - Decomposition	
Decomposition (ton/minute)	IF THEN ELSE (Dry Nitrocellulose <= 0, 0, IF THEN ELSE (Container Temperature < 40, 0, MIN(Dry Nitrocellulose, Container Temperature * Decomposition Rate)))	
Fire (unitless)	IF THEN ELSE (Container Temperature > Self Ignition Point, 1, 0)	
Container Temperature (degrees)	Initial Value = Room Temperature + 30	Temperature that day was 35 degrees, at which container temperature can reach 65 degrees [1].
	Equation = Decomposition Heat - Cooling	
Cooling (degree/minute)	IF THEN ELSE (Container Temperature > Room Temperature, Cooling System Activation, 0)	
Decomposition Heat (degree/minute)	Dry Nitrocellulose * Container Temperature * Decomposition Heat Rate	

References

1. State Council of the People's Republic of China Special Accident Investigation Team.: (Tianjin Port "8·12" Rui Hai International Logistic Company Dangerous Goods Storage Area Fire and Explosion Accident Report). http://www.chinasafety.gov.cn/newpage/Channel_21382.htm. Accessed 20 Apr 2018
2. Sklet, S.: Comparison of some selected methods for accident investigation. J. Hazard. Mater. **111**(1), 29–37 (2004)

3. Pasman, H.J., Rogers, W.J., Mannan, M.S.: How can we improve process hazard identification? What can accident investigation methods contribute and what other recent developments? A brief historical survey and a sketch of how to advance. J. Loss Prev. Process Ind. (2018)
4. Chakraborty, A., Ibrahim, A., Cruz, A.M.: A study of accident investigation methodologies applied to the Natech events during the 2011 great east japan earthquake. J. Loss Prev. Process Ind. **51** (2018)
5. Ferry, T.S.: Modern Accident Investigation and Analysis, 2nd edn. Wiley, New York (1988)
6. Leveson, N.: A new accident model for engineering safer systems. Saf. Sci. **42**, 237–270 (2004)
7. Cooke, D.L.: A system dynamics analysis of the Westray mine disaster. Syst. Dyn. Rev. **19** (2), 139–166 (2003)
8. Saleh, J.H., Marais, K.B., Bakolas, E., Cowlagi, R.V.: Highlights from the literature on accident causation and system safety: review of major ideas, recent contributions, and challenges. Reliab. Eng. Syst. Saf. **95**(11), 1105–1116 (2010)
9. DOE: Conducting Accident Investigations, DOE Workbook, Revision 2, US Department of Energy, Washington, DC, USA (1999)
10. Akter, S., Wamba, S.F.: Big data and disaster management: a systematic review and agenda for future research. Ann. Oper. Res. **9**, 1–21 (2017)
11. Huang, L., Wu, C., Wang, B., Ouyang, Q.: A new paradigm for accident investigation and analysis in the era of big data. Process Saf. Prog. **37**(1), 42–48 (2018)
12. Tirunagari, S.: Data mining of causal relations from text: analysing maritime accident investigation reports. Comput. Sci. (2015)
13. Sterman, J.D.: System dynamics modeling: tools for learning in a complex world. Calif. Manag. Rev. **43**(4), 8–25 (2001)
14. Zhou, A., Fan, L.: A new insight into the accident investigation: a case study of Tianjin Port fire and explosion in China. Process Saf. Prog. **36**(4), 362–367 (2017)

Generating Risk Maps for Evolution Analysis of Societal Risk Events

Nuo Xu[1,2] and Xijin Tang[1,2(✉)]

[1] Academy of Mathematics and Systems Science,
Chinese Academy of Sciences, Beijing 100190, China
xunuo1991@amss.ac.cn, xjtang@iss.ac.cn
[2] University of Chinese Academy of Sciences, Beijing 100049, China

Abstract. The development of societal risk events has been heavily concerned by both the government and the public. Faced with ever-increasing information, people struggle to follow the evolution of societal risk events. In order to identify the evolution of societal risk events, this paper presents an improved algorithm based on the method of generating information maps. One real-world case is illustrated and the evaluation is given. The improved approach for the evolution analysis whose results show the promising performance may be used for post-operation analysis, and decision-making process for government management.

Keywords: Risk maps · Evolution analysis · HNSW · Societal risk events

1 Introduction

In an era of information overload, it becomes increasingly difficult to keep up with the evolution of events. Although search engines are effective in retrieving information, people are easily lost in details and lose sight of the whole story. To better understand the complex stories, a structured story chain is expected to be generated automatically.

Societal risk events are those events involving major hazards and possible harm, raising the whole concern of society [1]. In Web 2.0 era, such events easily ignite widespread discussions and quickly turn into highlighted hot spots, then bring big challenges for government. Search engines have been the most common tools for the public to access information such as some social hot spots. Events that both attract most public attention and cause most search traffic are extracted and presented as Baidu hot news search words (HNSW), based on real-time search behaviors of Internet users and shown at Baidu News Portal by Baidu, the biggest Chinese search engine. Baidu HNSW as the user-generated contents actively reflect the public current concerns, and can be regarded as one kind of effective way to perceive societal risk. Therefore, Tang (2013) proposed to map HNSW into either risk-free events or one event with risk label from one of 7 risk categories including national security, economy/finance, public morals, daily life, social stability, government management, and resources/environment [2]; risk categories resulted from socio-psychological study [3]. A specific Web crawling system "HotWord Vision" was developed and improved

© Springer Nature Singapore Pte Ltd. 2018
J. Chen et al. (Eds.): KSS 2018, CCIS 949, pp. 115–128, 2018.
https://doi.org/10.1007/978-981-13-3149-7_9

to hourly download HNSW and their corresponding news texts since March of 2011 [4, 5]. With the HNSW list obtained every day, effective machine learning algorithms were explored to automatically label each HNSW either with risk-free category or one of 7 risk categories since November of 2011 [5, 6]. In addition, we manually verify the corresponding societal risk of those HNSW. The daily risk level with the frequency of daily risk events over the frequency of the daily HNSW was acquired. The effects of HNSW-based derived societal risk levels on China stock market volatility were studied using Granger causality analysis [7].

As HNSW correspond to societal risk events, the focus of this paper aims to explore the evolution of societal risk events and transitions of risks along the time by building comprehensive views to explicitly describe and capture public concerns. Various algorithms and models have been proposed for event evolution. Makkonen (2003) pioneered in setting event evolution as a sub-objective of topic detection and tracking (TDT) research [8]. Some event evolution studies with structured output [9, 10] focused mostly on timeline generation. However, this kind of summarization only worked for simple stories, which was linear in nature. In contrast, complex stories presented a very non-linear structure: stories with branches, side stories, and inter-twined narratives. Some researchers have moved to non-linear output and proposed different concepts of storylines. Wu et al. studied the event evolution based on random walk model by computing the importance of events which could happen in the next stage [11]. Zhang et al. constructed an event evolution model by using the events timestamp, events content similarity and events dependence between features [12]. Five different event evolution patterns were defined. Kalyanam et al. proposed a non-negative matrix factorization (NMF) based model which incorporated both the text content and social context to investigate topic involution [13]. Jia and Tang improved the maximum spanning tree algorithm on a multi-view graph to generate the storylines [14]. However, those above studies did not take the coherence of the storyline into consideration.

Shahaf et al. (2012) first proposed the methodology "metro maps" to explicitly capture story evolution and development [15]. Three criteria that characterized good metro maps were formalized. The objective function was built and optimized to generate storylines by considering all three criteria. The metro maps consisted of stops and lines. Each metro line was a sequence of metro stops and each metro stop was a single news article. Different lines focused on different aspects of the events. Shahaf et al. (2013) extended and scaled up the metro maps, with metro stops represented by sets of words (clusters) instead of single article [16]. However, the map generated by the algorithm "metro map" represented the evolution of topics by words, instead by events. In this paper, we improve the algorithm in [16] to increase the applicability for analysis of actual societal risk events from two perspectives. On one hand, the algorithm is extended from event-level with events represented by distributed vectors. On the other hand, different clustering methods are implemented to obtain set of events as the metro stops. We refer the event-level metro map as risk map.

This paper is organized as follows. The algorithm of constructing risk map for evolution of societal risk events is introduced in Sect. 2. Section 3 illustrates the development of risk map using one case "Chinese Red Cross". The comparisons of different algorithms are given. Conclusions and future work are given in Sect. 4.

2   Generating Risk Maps

In this section, the improved algorithm is addressed to show how societal risk events evolve and risks transfer along the time by better coherent storylines. The methodology "metro maps" proposed by Shahaf *et al.* creates structured summaries of stories by considering coherent chains of events generated by taking multiple criteria into account [16]. Given a large corpora of documents and a defined query, the definition of metro map is presented as follows.

Definition 1 (Metro Map). A metro map M is a pair (G, Π), where $G = (C; E)$ is a directed graph and Π is a set of paths in G. We refer paths as metro lines. Each $e \in E$ belong to at least one metro line.

Here, W is the vocabulary of corpora. Each document is a multi-set of W. Vertices C correspond to word clusters (subsets of W), and are denoted by stops (M). The lines of Π correspond to aspects of the story. The algorithm in [16] is designed to extract structured and coherent stories through connecting stops with multiple lines. Firstly, one key point is the representation of metro stops. Another is the generation of lines. As is known, the characteristic that a good storyline has is that following the clusters along a line tells a coherence story and gives a clear understanding of evolution of the event. A generative graph clustering method, BigClam, is applied to identify densely connected overlapping clusters as metro stops in [16]. The coherence is defined to measure the quality of a storyline. The similarity between each pair of consecutive clusters is calculated using BigClam. Moreover, "structure" and "coverage" are defined as two other characteristics which respectively control the number of coherent lines of a map and encourage the diversity of the story.

However, such an approach is inappropriate directly applied to evolution analysis of societal risk events and has two major shortcomings. Firstly, the processing for acquiring clusters is at word-level through grouping words that co-occur in one news article, and thus may lead fragmented and scattered topics. Secondly, the method of generating clusters based on overlapping cluster detection algorithm may bring redundant clusters, which leads to the incomplete coverage of story. As a result, we extend the metro map algorithm with processing from event-level rather than word-level. Word embedding model that learns vector representation for events is adopted to process clusters. The improvements not only capture semantic word relationships, but also help us understand relations among events. Furthermore, the construction of the map from event-level allows to see how the transitions of risks of events evolve along the time.

Our purpose is to generate the "risk map" which is represented by concise structured set of events, maximizing coverage of salient development of societal risk events. We refer the event-level metro map as "risk map". Similar to metro map, the risk map also consists of lines and stops. The "risk lines" are defined as links among sets of events in this paper. Each "risk line" is a sequence of stops which are defined as sets of societal risk events. Different lines focus on different developments of the events. Given societal risk events dataset S, the definition of "risk map" is defined as follows.

Definition 2 (Risk Map). A risk map R is a pair (G, Π), where $G = (C; E)$ is a directed graph and Π is a set of paths in G. We refer to paths as risk lines. Each $e \in E$ belong to at least one risk line.

Here, vertices C correspond to event clusters (subsets of S), denoted as stops. Three criteria "coherence", "connectivity" and "coverage", used in metro map, are revised for the risk map, and respectively elaborated and formalized as follows.

2.1 Coherence

The coherence of a "risk map" is defined to measure similarity between each pair of consecutive event clusters along one line. One piece of HNSW represents a societal risk event and the set of HNSW is divided into several time steps. The word embedding model Continuous Bag-of-Words (CBOW) as one of the model architectures proposed by Mikolov *et al.* is an efficient unsupervised algorithm for learning high-quality distributed vector representations which captures syntactic and semantic word relationships [17]. Societal risk events are represented by distributed vector representations by CBOW. As HNSW are generated by public searching behavior and consist of few words, the vector of one piece of HNSW are represented by averaging its word vectors. A variety of clustering methods are tried to obtain event clusters. Obviously, the quality of the cluster is essential for building a good map. The silhouette[1] is chosen as the quality measure of event clustering results. The silhouette ranges from -1 to $+1$, where a high value indicates that the object is well matched to its own cluster and poorly matched to neighboring clusters. In each time stamp, the best silhouette is selected to generate high quality clusters.

After formalizing cluster acquisition, we then turn to computing coherence of lines. To get coherence score, similarity between each pair of consecutive clusters along the line is computed. Here, Jaccard similarity coefficient is chosen as the measurement of coherence.

Given a line of clusters, we first score each transition by computing Jaccard similarity which is defined as the size of the intersection divided by the size of the union of the sets.

$$J\left(c_{i(j-1)}, c_{ij}\right) = \frac{\left|c_{i(j-1)} \cap c_{ij}\right|}{\left|c_{i(j-1)} \cup c_{ij}\right|}, j \in \{1, 2, \ldots, n\} \tag{1}$$

Here, c_{ij} denotes jth cluster in ith line, n denotes the number of clusters one line contains. Then the coherence of one line is calculated as:

$$line\_coh_i = \frac{\sum_{n-1} J(c_{i(j-1)}, j)}{n - 1} \tag{2}$$

---

[1] https://en.wikipedia.org/wiki/Silhouette_(clustering).

2.2   Coverage

In addition to coherence, high coverage is ensured for risk map. On one hand, the coverage of important aspects of the story is focused. On the other hand, diversity is expected. Therefore, we consider set-coverage as a sampling procedure: each line in the map tries to cover cluster c with probability $cover_l(c)$ which is defined as the proportion of clusters one line covers over all the clusters. The coverage of c is the probability at least one of the lines succeed:

$$cover_\Pi(c) = 1 - \prod_{l \in \Pi}(1 - cover_l(c)) \tag{3}$$

Thus, if the map includes lines which cover c well, $cover_\Pi(c)$ is close to 1. Adding another line which covers c well provides very little extra coverage of c, which encourages to pick lines which cover other clusters and promotes diversity.

2.3   Connectivity

The final criteria is connectivity. The goal of connectivity is to ensure the storylines intersect, rather than stand alone. There are several approaches to measure connectivity of a map: count the number of connected components, or perhaps the number of vertices that belong to more than one line. Here, we simply formalize connectivity as the number of lines of Π that intersect.

2.4   Algorithm

With three criteria of the risk map formally defined, we combine them into one objective function. We need to consider trade-offs among three criteria: for example, maximizing coherence often results in repetitive and low-coverage lines. Maximizing connectivity encourages choosing similar lines, resulting in low coverage as well. Maximizing coverage may cause low connectivity, leading to fewer intersecting lines. As a result, the appropriate objective function will be found to make trade-offs among the three criteria.

Suppose we pick connectivity as the primary objective. A stickier problem is that coherent lines tend to come in groups: a coherent line is often accompanied by multiple similar lines. Those lines intersect with each other. Choosing them will maximize connectivity. However, the map acquired will be highly redundant.

Because of that, we choose coverage as the primary objective. We first maximize coverage, next we maximize connectivity over maps that display maximal coverage. Let k be the maximal coverage across maps with *coherence* $\geq \tau$ (a given threshold). Now the problem is formulated as follows:

Problem. Given a set of candidate clusters C, find a map $R = (G, \Pi)$ over C which maximize connectivity, s.t. *Coherence*$(C) \geq \tau$ and *Cover*$(C) \geq k$.

To construct a good map, the first is to pick good lines. We expect to list all possible candidates. However, the number of possible lines may be exponential, and thus it is not applicable to enumerate them all. Restrict the number of lines is needed through restricting the map size. Add lines until coverage gains fall below a threshold and get the final number of lines K. The objective function is optimized that prefers longer coherent storylines whenever possible. For this, the greedy algorithm is adopted. At each round, compute approximate best-paths among clusters. Then greedily pick the best one amongst them for the map. The map obtained may show two forms, linear stories become linear maps; while complex stories maintain their intertwined threads.

Overall, the "risk map" operates as follows. Firstly, start with a large corpus of texts and a defined query (e.g., "Chinese Red Cross"). Secondly, all relevant texts are extracted and divided into several time steps. Thirdly, learn vector representations for texts and generate clusters in each time step. Finally, compute the scores of those three criteria respectively and optimize the objective function under constrained criteria as defined above.

3 Case Study

The events of "Chinese Red Cross" are chosen as a case study to conduct evolution analysis and transitions of societal risk along the time. In June of 2011, the name of "Guo Meimei" was very famous in Sina Weibo. The 20-year-old girl who showed off luxurious lifestyle claimed close relations with Chinese Red Cross. During that period, there emerged a large amount of HNSW about "Guo Meimei" and Chinese Red Cross, such as "The survey report of Guo Meimei" and "Chinese Red Cross rebuilt public trust". "Guo Meimei" incident exerted negative effects toward the donation to Chinese Red Cross. Compared with the donation after Wenchuan earthquake in May of 2008, the donation dropped sharply after Lushan earthquake in April of 2013. The foci of public opinions shifted from the case of "Guo Meimei" incident to the institution, governance, causing government trust crisis to some extent.

3.1 Data Processing

Baidu HNSW are provided in forms of 10 to 20 hot query news words updated every 5 min automatically which refer to bring the most search traffic. Each of HNSW corresponds to 1–20 news whose URLs are at the first page of hot words search results. As to "Chinese Red Cross" relevant events, we collect 82 HNSW including 633 news text and corresponding news text related to "Chinese Red Cross" from the cumulate corpus of HNSW downloaded since November of 2011. The time span of HNSW on "Chinese Red Cross" ranges from 2011 to 2016.

There are two strategies when dividing HNSW into chronological time steps: dividing HNSW into steps encompasses either a fixed amount of time or a fixed number of HNSW. The popularity of an event over time varies; there are bursts of intense popularity, followed by long periods of relative unpopularity. As we prefer to get more detail during popular times, we use the latter. Thus, divide the set of "Chinese Red Cross" related HNSW into 10 time steps. Then Chinese news text segmentation and stopword removal have been performed as preprocessing steps. To acquire more comprehensive semantic information, the news texts from April 1, 2013 to December 31, 2016 are chosen to learn word vector representations. The parameters are set as follows: the learnt vector representations are set to 100 dimensions, the optimal window size is 5. Furthermore, the top 20 keywords from news text are extracted as the contents of HNSW documents by TextRank [18]. The vector of one piece of HNSW is represented by averaging its word vectors. We respectively implement three clustering methods: K-means, hierarchical clustering and Gaussian mixtures algorithm. In each time stamp, select the best silhouette to generate high quality clusters. The best K and silhouette coefficient in each time step are shown in Fig. 1 by two lines as these two indexes are the same for K-means, hierarchical clustering and Gaussian mixtures algorithm.

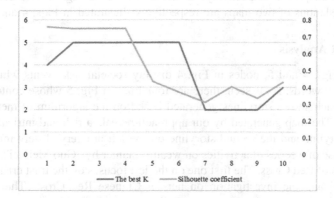

Fig. 1. The best K and silhouette coefficient

Once clusters are generated, we compute scores of "coherence", "coverage" and "connective" respectively. Then optimize the objective function to get risk map. The algorithm metro map is also applied to "Chinese Red Cross" dataset. The result comparisons between two algorithms are given next.

3.2 Evaluation

When we obtain the results, evaluating them quantitatively is a difficult task. There is no established golden standard. In this paper, we try to evaluate the results respectively from two aspects. One is from the quality of lines. The other is from the global structure of map. The average values of both coherence scores and length of lines are considered as the evaluation of line quality, while the coverage of map is considered as the evaluation of map structure. The average values of lines quality is calculated as Eq. (4):

$$aver\_line\_score = \frac{\sum_{l-1}(\mu_c * line\_coh_i + \mu_l * line\_len_i)}{l-1} \tag{4}$$

Here, $line\_len_i$ is the length of ith line, l is the number of lines, μ_c and μ_l are the weights. The values of both indicators in the case study are listed in Table 1.

Table 1. Comparisons of risk map and metro map

	Average of line quality	Map structure quality
Risk map	**0.36**	**0.42**
Metro map	0.34	0.20

As seen from the Table 1, our method outperforms metro map in both indicators, which has demonstrated its superiority for evolution analysis of social risk events with encouraging high coverage and diverse topics. The maps generated either by risk maps or by metro maps are respectively shown in Fig. 2 or Fig. 3. The corresponding English translations of two maps are respectively illustrated in Figs. 4 and 5.

3.3 Result Analysis

Compared Figs. 4 and 5, nodes in Fig. 4 display societal risk events while nodes in Fig. 5 display words. As is seen, there are four lines in Fig. 5 whose contents are not coherent, which shows storylines generated based on the algorithm of metro map are fragmented. The map generated by our approach reveals a rich and intricate structure. The first storyline and the second storyline on the left are very closely intertwined in Fig. 4 because of a series of activities on wealth flaunter by "Guo Meimei", which lead to trust crisis of Red Cross. The first one on the left focuses on the trust crisis caused by Guo Meimei and the investigation on her by Chinese Red Cross. There is a phenomenon that the consecutive stops contain the same event "Chinese Red Cross denied misappropriation of funds". This is because the events corresponding to HNSW appear on different days and are divided into two different time stamps. The second one on the left presents "Guo Meimei" incident, starting with flame war in Sina Weibo between Guo Meimei and Weng Tao who was the chairman of the associated enterprise of Chinese Red Cross and ending up with "Guo Meimei" arrested by the police. The third line describes the events about donation corruption on Wenchuan earthquake. In Fig. 5, the first one on the left tells about "Guo Meimei" incident. The second one mainly represents that the reputation of the Chinese Red Cross was damaged by "Guo Meimei". The line also contains donation corruption about Wenchuan earthquake. While the rest of two storylines respectively illustrate trust crisis of Chinese Red Cross and donation corruption about Wenchuan earthquake, these two storylines are incomplete and fragmented. The risk map generated by our algorithm clearly and comprehensively illustrates the evolution of societal risk events. Moreover, the risks of the events about wealth flaunter of "Guo Meimei" are public morality at the beginning. Later, the police found "Guo Meimei" suspected of gambling. The risks transfer from public morality to social stability. When we mentioned "Chinese Red Cross" before, the risks of related events are risk-free. But since the show-off incidents of Guo Meimei and donation corruption, Chinese Red Cross encountered public trust crisis with greatly donation decrease. The controversy was so heated that the Chinese Red Cross had to clarify it, and the investigation was carried out. The risks of events about Chinese Red Cross transfer from risk-free to government management.

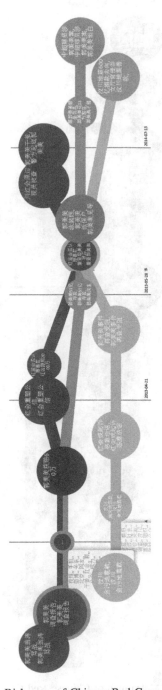

Fig. 2. Risk map of Chinese Red Cross events

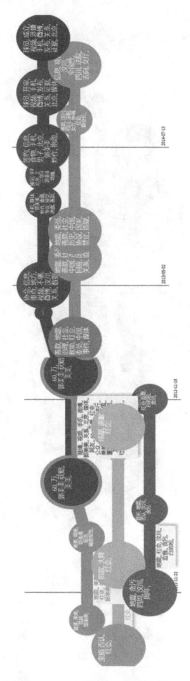

Fig. 3. Metro map of Chinese Red Cross events

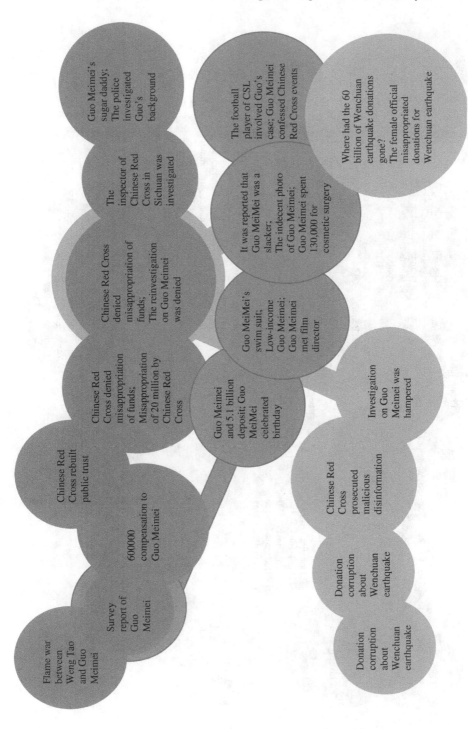

Fig. 4. English translation of risk map of Chinese Red Cross events

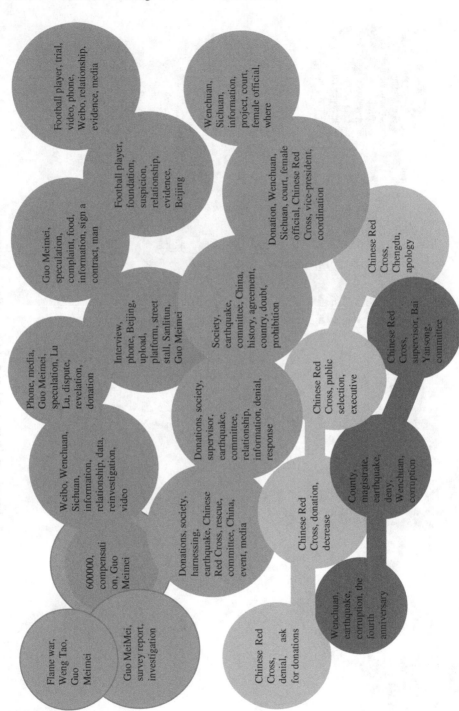

Fig. 5. English translation of metro map of Chinese Red Cross events

Overall, we conclude that our approach is more suitable for evolution analysis of societal risk events and explicitly shows the evolutionary relationships among HNSW in a way and captures the story development.

4 Conclusions

This paper focuses on exploring the evolution of societal risk events and transitions of risks along the time by building comprehensive views to explicitly describe and capture the information. We have improved the algorithm "metro maps" as the structured summaries of stories for the evolution analysis of societal risk events specially. Then, we test our algorithm on a real-world dataset. The study shows the promising performance of the proposed approach. The major contributions of this paper are summarized as follows:

(1) A general yet effective model, word embedding model, is introduced to learn vector representations for pieces of HNSW with adding semantic word relationships;

(2) The process of generating clusters for map construction is at the event-level rather than word-level, which allows to see not only how societal risk events evolve along the time, but also the transitions of risks.

The structured maps extracted from Baidu HNSW derived online public concerns show a more distinct and comprehensive picture of development of societal risk events. It is of great help to understand the situation when mapping out how risk events evolve clearly and accurately. An appropriate intervention could be taken at crucial stages for government administration in case things get worse. Lots of work needs to be improved. More effective clustering methods will be explored and real-world cases will be conducted in the future.

Acknowledgement. This research is supported by National Key Research and Development Program of China (2016YFB1000902) and National Natural Science Foundation of China (61473284 & 71731002).

References

1. Ball, D.J., Boehmer-Christiansen, S.: Societal concerns and risk decisions. J. Hazard. Mater. **144**, 556–563 (2007)
2. Tang, X.J.: Exploring on-line societal risk perception for harmonious society measurement. J. Syst. Sci. Syst. Eng. **22**(4), 469–486 (2013)
3. Zheng, R., Shi, K., Li, S.: The influence factors and mechanism of societal risk perception. In: Zhou, J. (ed.) Complex 2009. LNICST, vol. 5, pp. 2266–2275. Springer, Heidelberg (2009)
4. Wu, D., Tang, X.J.: Preliminary analysis of baidu hot words. In: Proceedings of the 11th Youth Conference on Systems Science and Management Science, pp. 478–483 (2011). (in Chinese)

5. Hu, Y., Tang, X.J.: Using support vector machine for classification of baidu hot word. In: Wang, M. (ed.) KSEM 2013. LNCS, vol. 8041, pp. 580–590. Springer, Heidelberg (2013)
6. Xu, N., Tang, X.J.: Exploring effective methods for on-line societal risk classification and feature mining. In: Cheng, X.Q., et al. (eds.) Chinese National Conference on Social Media Processing. CCIS, vol. 774, pp. 65–76. Springer, Heidelberg (2017)
7. Xu, N., Tang, X.J.: Societal risk and stock market volatility in china: a causality analysis. In: Chen, J., et al. (eds.) The 18th International Symposium on Knowledge and Systems Sciences. CCIS, vol. 780, pp. 175–185. Springer, Heidelberg (2017)
8. Makkonen, J.: Investigations on event evolution in TDT. In: Proceedings of HLT-NAACL Student Research Workshop, vol. 3, pp. 43–48. ACL (2003)
9. Allan, J., Gupta, R., Khandelwal, V.: Temporal summaries of new topics. In: Proceedings of the 24th Annual International ACM SIGIR Conference on Research and Development in Information Retrieval. vol. 23, pp. 10–18. ACM (2001)
10. Yan, R., et al.: Evolutionary timeline summarization: a balanced optimization framework via iterative substitution. In: Proceedings of the 34th International ACM SIGIR Conference on Research and Development in Information Retrieval, pp. 745–754. ACM (2011)
11. Wu, C., Wu, B., Wang, B.: Event evolution model based on random walk model with hot topic extraction. In: Li, J.Y., et al. (eds.) ADMA 2016. LNAI, vol. 10086, pp. 591–603. Springer, Heidelberg (2016)
12. Kalyanam, J., et al.: Leveraging social context for modeling topic evolution. In: Proceedings of the 21st ACM SIGKDD International Conference on Knowledge Discovery and Data Mining, pp. 517–526. ACM (2015)
13. Zhang, H., et al.: Modeling news event evolution. J. Natl. Univ. Def. Technol. 35(4), 166–170 (2013). (in Chinese)
14. Jia, Y.G., Tang, X.J.: Generating storyline with societal risk from tianya club. J. Syst. Sci. Inf. 5(6), 524–536 (2017)
15. Shahaf, D., Guestrin, C., Horvitz, E.: Trains of thought: Generating information maps. In: Proceedings of the 21st International Conference on World Wide Web, pp. 899–908. ACM (2012)
16. Shahaf, D., et al.: Information cartography: creating zoomable, large-scale maps of information. In: Proceedings of the 19th ACM SIGKDD International Conference on Knowledge Discovery and Data Mining, pp. 1097–1105. ACM (2013)
17. Mikolov, T., et al.: Efficient estimation of word representations in vector space. arXiv:1301.3781v3[cs.CL], pp. 1–12 (2013)
18. Mihalcea, R., Tarau, P.: TextRank: bringing order into texts. In: Proceedings of Empirical Methods on Natural Language Processing, pp. 404–411. ACL (2004)

Research on Forest Fire Processing Scheme Generation Method Based on Belief Rule-Base

Yan Xu$^{(\boxtimes)}$, Ning Wang, Xuehua Wang, Zijian Ni, Huaiming Li,
and Xuelong Chen

Faculty of Management and Economics, Dalian University of Technology,
Dalian 116024, China
120712416@qq.com

Abstract. In order to gain better experience and knowledge from historical disaster response and improve the ability of decision support in emergency management, a method of mining processing schemes based on belief rule-base is proposed based on forest fire history cases. Combined with the historical data of American forest fires in 2014 provided by the National Fire Incident Reporting System (NFIRS), the data extraction of fire status and coping strategies was realized, and the rules were mined by Apriori algorithm to form the forest fire processing rule-base. The reasoning model of belief rule-base for practical business is constructed to realize the optimal selection of fire processing scheme and to provide a supporting scheme for forest fire response decision.

Keywords: Belief Rule-Base · RIMER · Forest fire · Emergency management

1 Introduction

The occurrence and spread of forest fire has brought inestimable economic loss and deterioration of ecological environment to the country and society, and even caused serious casualties. In order to deal with forest fire better, avoid forest fire as much as possible and reduce the loss caused by forest fire, how to make scientific and effective decision and respond quickly is the key to solve the problem.

In recent years, there have been a lot of researches on fire prediction and decision making by domestic and foreign scholars, but most of them have been studied from the perspective of fire evolution, For example, Li et al. [1] built a Bayesian Network (BN) evolution model of high-rise building fire, and use joint probability formula to predict fire consequences; Wang [2] used Bayesian Network method to probe into the evolution of oil pipeline leakage fire explosion accident. However, there are relatively few researches on coping based on human behavior; Liu [3] made use of various machine learning techniques, such as support vector machine (SVM) and random forest, to analyze their characteristic indexes and predict forest fire area, so as to improve the management and allocation of fire protection resources; Cai et al. [4] realized forest fire behavior prediction, simulated and generated decision plan by using

© Springer Nature Singapore Pte Ltd. 2018
J. Chen et al. (Eds.): KSS 2018, CCIS 949, pp. 129–141, 2018.
https://doi.org/10.1007/978-981-13-3149-7_10

ArcInfo technology, which is beneficial to forest fire command and rescue decision; Li [5] used least squares support vector machine (LS-SVM) to build a model to excavate the knowledge in forest fire cases and predict the burned forest area, but lack of solutions to fire emergencies.

Because of the uncertainty of sudden fire events, RIMER(Belief rule-base inference methodology using the evidential reasoning approach) is an effective method to deal with all kinds of uncertain data. RIMER has been widely used in many fields because of its ability to deal with all kinds of uncertainties. For example, Wang et al. [6] proposed a hybrid evidence-based reasoning (ER) and confidence rule base (BRB) method for consumer preference prediction, and applied it to orange juice; Xu et al. [7] applied belief rule-base to pipeline leak detection, and proposed a method of training and detection based on belief rule-base. It has been proved that the leak detection system based on belief rule-base has strong flexibility. It is an effective new method for pipeline leakage detection; Kong et al. [8] proposed a decision support system for clinical risk assessment of cardiothoracic pain based on belief rule-base, which is used to deal with the uncertainty of clinical knowledge and clinical data, and achieved good results.

To sum up, the fire response decision needs scientific methods to support, and RIMER method has certain value to the causal decision problem of uncertainty. Therefore, in this paper, the theory of belief rule-base, modeling method and its application in practical problems are deeply analyzed, combined with the historical data series of American forest fires in 2014 provided by the National Fire Incident Reporting System (NFIRS). The inference model of belief rule-base is constructed by using the method of emergency processing scheme generation based on belief rule-base to realize the optimal selection of fire processing scheme.

2 Theories

2.1 Knowledge Representation of Belief Rule-Base

The methods commonly used in uncertain reasoning include credibility method [9], subjective Bayes method [10], evidence theory [11], etc. Belief rule-base (BRB) consists of a series of if-then rules with belief degree. The kth belief rule of BRBB is described as follows,

$$R_k : If\ A_1^k \bigwedge A_2^k \bigwedge \cdots \bigwedge A_{M_k}^k, Then\{(D_1, \beta_{1,k}), (D_2, \beta_{2,k}), \dots, (D_N, \beta_{N,k})\}$$
$$\text{with a rule weight } \theta_k \text{ and attribute weights } \delta_{1,k}, \delta_{2,k}, \cdots, \delta_{M_k,k} \tag{1}$$

where $A_i^k(i = 1, 2, \cdots, M_k, k = 1, 2, \cdots, L)$ is the referential value of the ith antecedent attribute in the kth rule, M_k the number of antecedent attributes used in the kth rule; L the number of rules in BRB, $\theta_k(k = 1, 2, \cdots, L)$ the relative weight of the kth rule, $\delta_{i,k}(i = 1, 2, \cdots, M_k, k = 1, 2, \cdots, L)$ are the relative weights of the M_k antecedent attributes used in the kth rule, $\beta_{j,k}(j = 1, 2, \cdots, N, k = 1, 2, \cdots, L)$ the belief degree to which D_j is believed to be the consequent in the kth rule. If $\sum_{j=1}^{N} \beta_{j,k} = 1$, then the kth

packet rule is said to be complete; otherwise, it is incomplete.

A BRB consisting of L belief rules can also be expressed as a belief structure such as that shown in Table 1.

Table 1. Belief rule base

Rule	Rule weight	Attribute weight (Input)				Consequent (Output)			
		$A_1(\delta_1)$	$A_2(\delta_2)$	$\cdots$	$A_M(\delta_M)$	D_1	D_2	$\cdots$	D_N
1	θ_1	$A_1^1(\delta_{1,1})$	$A_2^1(\delta_{2,1})$	$\cdots$	$A_M^1(\delta_{M,1})$	$\beta_{1,1}$	$\beta_{2,1}$	$\cdots$	$\beta_{N,1}$
2	θ_2	$A_1^2(\delta_{1,2})$	$A_2^2(\delta_{2,2})$	$\cdots$	$A_1^2(\delta_{M,2})$	$\beta_{1,2}$	$\beta_{2,2}$	$\cdots$	$\beta_{N,2}$
$\vdots$	$\vdots$	$\vdots$	$\vdots$	$\cdots$	$\vdots$	$\vdots$	$\vdots$	$\cdots$	$\vdots$
L	θ_L	$A_1^L(\delta_{1,L})$	$A_2^L(\delta_{2,L})$	$\cdots$	$A_M^L(\delta_{M,L})$	$\beta_{1,L}$	$\beta_{2,L}$	$\cdots$	$\beta_{N,L}$

2.2 RIMER Approach

The basic idea of RIMER is that when the input information x arrives, Using ER algorithm to combine the belief rules in BRB to obtain the final output of BRB system. RIMER mainly realizes the inference of BRB system by calculating the activation weight and synthesizing the activation rules [12].

Activation Weight Calculation. The activation weight of input information x to rule k can be calculated by formula (2),

$$\omega_k = \frac{\theta_k \prod_{i=1}^{M} (\alpha_i^k)^{\bar{\delta}_i}}{\sum_{l=1}^{L} \theta_l \prod_{i=1}^{M} (\alpha_i^l)^{\bar{\delta}_i}} \tag{2}$$

where $\omega_k \in [0,1]$, k $= 1, 2, \cdots, $L; $\delta_i = \frac{\delta_i}{\max_{i=1,2,\cdots,M}\{\delta_i\}}$; $\alpha_i^k (i = 1, 2, \cdots, M)$ represents the belief degree of the ith input relative to the reference value A_i^k in kth rule. When $\omega_k = 0$, it means that the kth rule is not activated; otherwise, the rule is activated.

Activation Rule Synthesis. Use the evidential reasoning algorithm to the activation rules so that they can be synthesized

On the basis of ER iterative algorithm, Wang and Yang et al. proposed evidence reasoning analysis algorithm [13]. By combining all the rules in BRB, the final output $S(x)$ of BRB can be obtained as follows,

$$S(x) = \left\{ (D_j, \hat{\beta}_j), j = 1, 2, \cdots, N \right\} \tag{3}$$

and

$$\hat{\beta}_j = \frac{\mu \times \left[\prod_{k=1}^{L} \left(\omega_k \beta_{j,k} + 1 - \omega_k \sum_{i=1}^{N} \beta_{i,k} \right) - \prod_{k=1}^{L} \left(1 - \omega_k \sum_{i=1}^{N} \beta_{i,k} \right) \right]}{1 - \mu \times \left[\prod_{k=1}^{L} (1 - \omega_k) \right]},$$

$$\mu = \left[\sum_{j=1}^{N} \prod_{k=1}^{L} \left(\omega_k \beta_{j,k} + 1 - \omega_k \sum_{i=1}^{N} \beta_{i,k} \right) - (N-1) \prod_{k=1}^{L} \left(1 - \omega_k \sum_{i=1}^{N} \beta_{i,k} \right) \right]^{-1} \tag{4}$$

where $\hat{\beta}_j$ is the function of $\bar{\delta}_i (i = 1, 2, \cdots, M)$, $\theta_k (k = 1, 2, \cdots, L)$ and $\beta_{j,k} (j = 1, 2, \cdots, N, k = 1, 2, \cdots, L)$.

3   A Model of Emergency Processing Scheme Generation Based on Belief Rule-Base

A case is usually represented by two parts, namely, the description of the status of the emergency and the description of the processing scheme of the emergency [14]. Therefore, this paper uses the form of binary group to represent the case, that is, case = (status, scheme).

The problem to be solved in this paper is to create an effective emergency processing scheme for the target case based on the status and scheme of historical case set. In order to generate a scientific and reasonable emergency processing scheme based on the historical case set, a method of emergency processing scheme generation based on the belief rule-base is applied.

By constructing the historical case rule base and using the reasoning method based on the belief rule-base, the solution of the target case can be obtained, and an emergency processing scheme for the target case can be generated. The detailed calculation steps of its generating method are as follows:

Step1: Transforming status attribute of historical case into belief distribution form $S(x_i) = \left\{ (A_{i,j}, \alpha_{i,j}), i = 1, 2, \cdots, M, j = 1, 2, \cdots, J_i \right\}$ by using Information Transformation Method [9]. Assume that the ith input information of the brb system $x_i (i = 1, 2, \cdots, M)$ is in the form of a quantitative value, when the information is transformed, it can be converted into the belief degree of each input data corresponding to the reference values on both sides. First of all, a rule is established by the decision maker to establish a corresponding relationship between the reference value $A_{i,j} (j = 1, 2, \cdots, J_i)$ and the numerical value $\gamma_{i,j}$:

$$\gamma_{i,j} \text{ means } A_{i,j} \, i = 1, 2, \cdots, M, j = 1, 2, \cdots, J_i$$

Assume that the value of x_i is between $\gamma_{i,j}$ and $\gamma_{i,j+1} (j = 1, 2, \cdots, J_i - 1)$ determined by the decision maker, then the belief degree of these two reference values $\alpha_{i,j}$ and $\alpha_{i,j+1}$ corresponding to x_i can be calculated by the following formula.

$$\begin{cases} \alpha_{i,j} = \dfrac{\gamma_{i,j+1} - x_i}{\gamma_{i,j+1} - \gamma_{i,j}}, \gamma_{i,j} \le x_i \le \gamma_{i,j+1}, j = 1,2,\cdots,J_i - 1 \\[3mm] \alpha_{i,j+1} = 1 - \alpha_{i,j}, \gamma_{i,j} \le x_i \le \gamma_{i,j+1}, j = 1,2,\cdots,J_i - 1 \\[3mm] \alpha_{i,s} = 0, s = 1,2,\cdots,J_i, s \ne j,j+1 \end{cases} \tag{5}$$

Then x_i can be translated into the following equivalent expectation form,

$$S(x_i) = \left\{ (A_{i,j}, \alpha_{i,j}), i = 1,2,\cdots,Mj = 1,2,\cdots,J_i \right\}$$

Step2:Determine three parameters $(\theta_k, \delta_{i,k}, \beta_{j,k})$ in BRB by historical case data, then train and adjust constantly. Yang et al. [15] proposed a BRB parameter training model in 2007, as shown by Fig. 1.

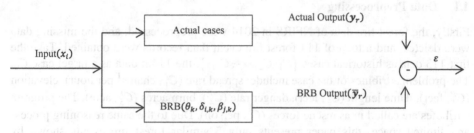

Fig. 1. BRB parameter training model.

Wang et al. [16] In the process of predicting consumer preference, the optimization model of parameter estimation is constructed. Based on this, the following optimization models are obtained:

$$\text{Minimize } J = \sum_{r=1}^{R} \varepsilon^2 = \sum_{r=1}^{R}(y_{rb} - \bar{y}_{rb})^2 \quad\text{.........①}$$
$$\text{s.t.} \begin{cases} \sum_{j=1}^{N} \beta_{j,k} = 1, k = 1,2,\cdots,L \ \text{......②} \\[2mm] \sum_{i=1}^{M} \delta_{i,k} = 1 \quad\text{.......................③} \\[2mm] 0 \le \beta_{j,k} \le 1 \quad\text{.......................④} \\[2mm] \delta_{i,k} \ge 0 \quad\text{.............................⑤} \\[2mm] 0 \le \theta_k \le 1, k = 1,2,\cdots,L \ \text{.........⑥} \end{cases} \tag{6}$$

Where formula ① is the objective function to make the deviation between the actual output value and the predicted output value as small as possible, formula ② and formula ④ is the constraint on the belief degree of the result of each belief rule, formula ③ and formula ⑤ is the constraint on the weight of the premise attribute. Formula ⑥ is a constraint on the weight of a rule.

Step3:By using formula (5), the status of the target case is converted into the form of belief distribution.

Step4:In order to obtain the rules activated by the target case, the activation weight ω_k of the target case is calculated.

Step5:The conclusion belief degree of target case is obtained by rule synthesis based on evidence reasoning, and the result is evaluated based on utility.

Step6:Repeat the above steps to obtain additional properties of the processing scheme.

4 Experiment

Experimental data comes from the forest fire data of America in 2014 of the National Fire Incident Reporting System (NFIRS) database. The main flow of the experiment includes knowledge representation, rule mining, belief rule-base construction and RIMER reasoning. The reasoning result is compared with the actual situation by the above method.

4.1 Data Preprocessing

Firstly, the forest fire data of NFIRS in 2014 were preprocessed, and the missing data were deleted, and a total of 114 forest fire event data records were obtained. Take the first 113 data as historical cases $\{C_1, C_2, \cdots, C_{113}\}$, the 114th data as a target case C_0. The problem attributes of the case include: spread rate (C_1^P, chains[1] per hour), elevation (C_2^P, feet), flame length (C_3^P, feet), danger rate (C_4^P), burn acres (C_5^P, acre). The program attributes are called in as rescue forces (C_1^S, person). Due to the same reasoning process and limited space, this paper presents only 5 similar forest fire events shown by Table 2.

Table 2. The historical data of the case base

No.	C_1^P	C_2^P	C_3^P	C_4^P	C_5^P	C_1^S
1	5	7000	4	3	1	6
2	1	1000	4	4	1	4
3	120	520	3	3	13	40
4	66	4000	10	1	20	34
5	1	236	1	2	4	13
...	...	...	...	...	...	...

In order to avoid producing too many rules, each attribute is divided into evaluation grades, and the criterion of reference value selection is not unique. The commonly used numerical classification methods include k-fold cross-validation, leave one group out method and random partitioning method, and so on. In this paper, the spread rate, elevation, flame length, burn acres and rescue forces are set to 4 levels,the reference value of it is set from low to high as $A = \{1, 2, 3, 4\}$, where spread rate is divided into

---

[1] Chain: A chain is 66 feet (20.1 m).

$(0,30]$, $(30,90]$, $(90,150]$, $(150,inf)$ chain per hour; elevation is divided into $(0,1000]$, $(1000,3500]$, $(3500,6000]$, $(6000,inf)$ feet; flame length is divided into $(0,8]$, $(8,28]$, $(28,48]$, $(48,inf)$ feet;burn acres is divided into $(0,5]$, $(5,50]$, $(50,5000]$, $(5000,inf)$ acre; rescue forces is divided into $(0,50]$, $(50,500]$, $(500,1500]$, $(1500,inf)$ person. The danger rate are set to 5 levels, the reference value of it is set as $A = \{1, 2, 3, 4, 5\}$, represents {low, medium, high, very high, extremely high}, thus the conclusion is drawn as Table 3.

Table 3. Reference evaluation grade

No.	A_1	A_2	A_3	A_4	A_5	D
1	1	4	1	3	1	1
2	1	1	1	4	1	1
3	3	1	1	3	2	1
4	2	3	2	1	2	1
5	1	1	1	2	1	1
...	...	...	...	...	...	...

4.2 Rule Mining

The data set is saved as csv format by Excel and imported into Weka software. The data is preprocessed by Filter as a non-numeric type, and the rules are mined by Apriori algorithm.

The computational results of the software are outputted in the following form of rules

$$\text{Best rules found:}$$

$$A_1 = A_1^1 A_2 = A_2^1 p \Rightarrow D = D_1 q \tag{6}$$

$$A_1 = A_1^2 D = D_2 p \Rightarrow A_2 = A_2^2 q \tag{7}$$

Suppose the software co-imports the set of S data items, the meaning of formula (6) is: There are p items in the data item set including $A_1 = A_1^1$ and $A_2 = A_2^1$, there are q items in these data items including $D = D_1$, then the belief degree of this rule is conf = c = q/p, the support degree is sup = p/S, where conf is not less than the value of minMetric and sup is not less than the value of lowerBoundMinSupport. Formula (7) dissatisfies the rule condition of status to scheme, so eliminate non - conforming rules from the result.

Filter the results according to the filtering conditions, If the last term of the rule is rescue forces, the preceding term is a non-empty subset of the set of premises attributes (that is, not including rescue forces), then it is retained, otherwise deleted. Examine one by one in this way, and get the filtered partial rule examples as follows:

14. elevation=1 danger rate=2 39 ==> rescue forces=1 39   <conf:(1)>

19. spread rate=1 elevation=1 danger rate=2 39 ==> rescue forces=1 39 <conf:(1)>

22. elevation=1 flame length=1 danger rate=2 39 ==> rescue forces=1 39 <conf:(1)>

26. spread rate=1 elevation=1 flame length=1 danger rate=2 39 ==> rescue forces=1 39   <conf:(1)>

33. danger rate=2 burn acres=1 27 ==> rescue forces=1 27   <conf:(1)>

...   ...

Then, merge rules and come to such a rule as: $if\ A_1 = A_1^1 A_2 = A_2^1$, $then\{(D_1, c_1), (D_2, c_2), \cdots, (D_N, c_N)\}$. It can be explained that each data item is regarded as a rule, that is, a total of S rules, and the belief degree relative to the rating D_j is $\beta_{j,k} = q/p = c_j$; The proportion of support for this rule reflects the importance of rules, that is, the rule weight $\theta_k = p/S = \text{sup}$. The combined partial rules are shown as follows:

If spread rate = 1 elevation = 1 flame length = 1 danger rate = 2 burn acres = 1 then rescue forces = {(1,1)}, $\theta_k = 24/114 = 0.21$;

If spread rate = 1 elevation = 1 flame length = 1 danger rate = 2 burn acres = 2 then rescue forces = {(1,1)}, $\theta_k = 14/114 = 0.12$;

If spread rate = 1 elevation = 2 flame length = 1 danger rate = 3 burn acres = 1 then rescue forces = {(1,1)}, $\theta_k = 7/114 = 0.06$;

If spread rate = 1 elevation = 1 flame length = 1 danger rate = 1 burn acres = 1 then rescue forces = {(1,1)}, $\theta_k = 5/114 = 0.04$;

If spread rate = 1 elevation = 1 flame length = 1 danger rate = 3 burn acres = 2 then rescue forces = {(1,1)}, $\theta_k = 4/114 = 0.04$;

If spread rate = 4 elevation = 2 flame length = 2 danger rate = 4 burn acres = 4 then rescue forces = {(4,1)}, $\theta_k = 1/114 = 0.01$;

If spread rate = 1 elevation = 1 flame length = 1 danger rate = 3 burn acres = 1 then rescue forces = {(1,0.93),(2,0.07)}, $\theta_k = 14/114 = 0.12$;

If spread rate = 1 elevation = 1 flame length = 1 danger rate = 3 burn acres = 3 then rescue forces = {(1,0.5),(2,0.5)}, $\theta_k = 2/114 = 0.02$;

If spread rate = 1 elevation = 2 flame length = 1 danger rate = 3 burn acres = 2 then rescue forces = {(1,0.5),(2,0.5)}, $\theta_k = 2/114 = 0.02$;
. . . . . .

Therefore, the rules are mined from the data $R_k : If A_1^k \bigwedge A_2^k \bigwedge \cdots \bigwedge A_{M_k}^k, Then\{(D_1, \beta_{1,k}), (D_2, \beta_{2,k}), \ldots, (D_N, \beta_{N,k})\}$ with rule weight θ_k. Finally, we get the set of relevant belief rules as shown in Table 4.

Table 4. Set of related belief rules.

No.	Rule weight	Antecedent attributes	Consequent
1	0.21	spread rate = 1 ∧ elevation = 1 ∧ flame length = 1 ∧ danger rate = 2 ∧ burn acres = 1	rescue forces = {(1,1)}
2	0.12	spread rate = 1 ∧ elevation = 1 ∧ flame length = 1 ∧ danger rate = 2 ∧ burn acres = 2	rescue forces = {(1,1)}
3	0.12	spread rate = 1 ∧ elevation = 1 ∧ flame length = 1 ∧ danger rate = 3 ∧ burn acres = 1	rescue forces = {(1,0.93), (2,0.07)}
4	0.06	spread rate = 1 ∧ elevation = 2 ∧ flame length = 1 ∧ danger rate = 3 ∧ burn acres = 1	rescue forces = {(1,1)}
5	0.04	spread rate = 1 ∧ elevation = 1 ∧ flame length = 1 ∧ danger rate = 1 ∧ burn acres = 1	rescue forces = {(1,1)}
6	0.04	spread rate = 1 ∧ elevation = 1 ∧ flame length = 1 ∧ danger rate = 3 ∧ burn acres = 2	rescue forces = {(1,1)}
7	0.02	spread rate = 1 ∧ elevation = 1 ∧ flame length = 1 ∧ danger rate = 3 ∧ burn acres = 3	rescue forces = {(1,0.5), (2,0.5)}
8	0.02	spread rate = 1 ∧ elevation = 2 ∧ flame length = 1 ∧ danger rate = 3 ∧ burn acres = 2	rescue forces = {(1,0.5), (2,0.5)}
9	0.01	spread rate = 4 ∧ elevation = 2 ∧ flame length = 2 ∧ danger rate = 4 ∧ burn acres = 4	rescue forces = {(4,1)}
...	...	...	...

Table 5. Antecedent attribute result

	Spread rate (x_1)	Elevation (x_2)	Flame length (x_3)	Danger rate (x_4)	Burn acres (x_5)
δ_i	0.13	0.29	0.28	0.16	0.14

4.3   Establishment of Belief Rule-Base

First, calculate the weight of the antecedent attribute. In this paper, the principal component analysis method is used to calculate the weight of the antecedent attribute, and the result of the weight of the antecedent attribute is obtained as shown by Table 5 (preserving two decimal places):

Finally, build the belief rule base as shown in Table 6.

Table 6. Belief rule base

Rule	Rule weight	Antecedent attribute (weight)					Consequent (belief degree)			
		Spread rate (0.13)	Elevation (0.29)	Flame length (0.28)	Danger rate (0.16)	Burn acres (0.14)	Rescue forces			
							1	2	3	4
1	0.21	1	1	1	2	1	1	0	0	0
2	0.12	1	1	1	2	2	1	0	0	0
3	0.12	1	1	1	3	1	0.93	0.07	0	0
4	0.06	1	2	1	3	1	1	0	0	0
5	0.04	1	1	1	1	1	1	0	0	0
⋮	⋮	⋮	⋮	⋮	⋮	⋮	⋮	⋮	⋮	⋮
39	0.04	1	1	1	3	2	1	0	0	0
40	0.02	1	1	1	3	3	0.5	0.5	0	0
41	0.02	1	2	1	3	2	0.5	0.5	0	0
42	0.01	4	2	2	4	4	0	0	0	1

4.4 RIMER Reasoning

Step1: According to the formula (5), the antecedent attribute of historical cases are converted to belief degree form, and the first fire event recorded data is taken as an example (Table 7):

Table 7. Spread rate (x_1).

$A_{i,j}$	1	2	3	4
$\gamma_{i,j}$	0	30	90	150

The conversion results are as follows: $S(x_1) = \{(1, 0.83), (2, 0.17), (3, 0), (4, 0)\}$. That means the spread rate of the fire incident is in Level 1, the belief degree of Level 1 is 0.83 and the belief degree of Level 2 is 0.17 (Table 8).

Table 8. Elevation (x_2).

$A_{i,j}$	1	2	3	4
$\gamma_{i,j}$	0	1000	3500	6000

The conversion results are as follows: $S(x_2) = \{(1, 0), (2, 0), (3, 0), (4, 1)\}$. That means the elevation of the fire incident is in Level 4 (Table 9).

Table 9. Flame length (x_3).

$A_{i,j}$	1	2	3	4
$\gamma_{i,j}$	0	8	28	48

The conversion results are as follows: $S(x_3) = \{(1, 0.5), (2, 0.5), (3, 0), (4, 1)\}$. That means the flame length of the fire incident is in Level 4 (Table 10).

Table 10. Danger rate (x_4).

$A_{i,j}$	1	2	3	4	5
$\gamma_{i,j}$	1	2	3	4	5

The conversion results are as follows: $S(x_4) = \{(1, 0), (2, 0), (3, 1), (4, 0), (5, 0)\}$. That means the danger rate of the fire incident is in Level 3 (Table 11).

Table 11. Burn acres (x_5).

$A_{i,j}$	1	2	3	4
$\gamma_{i,j}$	0	5	50	5000

The conversion results are as follows: $S(x_5) = \{(1, 0.8), (2, 0.2), (3, 0), (4, 0)\}$. That means the burn acres of the fire incident is in Level 1.

Step2: Establish the belief rule base and determine the parameters $(\theta_k, \delta_{i,k}, \beta_{j,k})$, the result is as shown in Table 6.

Step3: Convert the antecedent attribute of the target case to belief degree form as follows

$$S(x_1) = \{(1, 0.67), (2, 0.33), (3, 0), (4, 0)\}$$
$$S(x_2) = \{(1, 0.20), (2, 0.80), (3, 0), (4, 0)\}$$
$$S(x_3) = \{(1, 0.88), (2, 0.12), (3, 0), (4, 0)\}$$
$$S(x_4) = \{(1, 0), (2, 1), (3, 0), (4, 0), (5, 0)\}$$
$$S(x_5) = \{(1, 0.8), (2, 0.2), (3, 0), (4, 0)\}$$

Step4: Calculate the activation weight of the target case ω_k, thus the rules activated by the events of the target case in the belief rule base are obtained according to ω_k. $\omega_1 = 0.56$, $\omega_2 = 0.16$, $\omega_{10} = 0.22$, $\omega_{20} = 0.06$, the activation weight of other rules is 0. That means only the first, the second, the 10th and the 20th rules are activated, while other rules are not activated.

Step5: The rules are synthesized by ER to obtain the belief degree of the target case, and the results are evaluated based on utility. According to formula (4), it is calculated

$\mu = 1.04$, $\hat{\beta}_1 = 0.97$, $\hat{\beta}_2 = 0.03$, $\hat{\beta}_3 = \hat{\beta}_4 = 0$. The output which is the rescue forces of the final BRB in the form of formula (3) is as follows:

$$S(x) = \{(1, 0.97), (2, 0.03), (3, 0), (4, 0)\}$$

The result shows that,the belief degree of Level 1 is 0.97 and the belief degree of Level 2 is 0.03, so the decision scheme of rescue forces is level 1. By comparison, the reasoning result is consistent with the actual result.

5   Conclusions

In this paper, the belief rule base is established by using the processing scheme generation model based on the belief rule base, and the emergency processing scheme of the target case is obtained by the belief rule base reasoning method based on evidence reasoning. The belief rule base can not only deal with fire emergencies with uncertain information, but also generate more effective emergency processing schemes for target cases. In the example analysis of forest fire incident, the feasibility of applying belief rule base to fire incident emergency disposal is analyzed, and the optimal fire processing scheme is realized at the same time. It is helpful to provide decision support for decision-makers in the field of emergency management.

Acknowledgments. This work was supported by the National Natural Science Foundation of China (No. 71774021, No. 71373034, No. 71533001).

References

1. Li, J.X., Xia, D.Y., Wu, X.P.: Prediction model of fire consequence of high-rise building based on Bayesian network. China Saf. Sci. J. **23**(12), 54–59 (2013)
2. Wang, Q.Q.: Accident evolution and emergency evacuation analysis of oil pipeline leakage fire and explosion. China Saf. Sci. J. **26**(5), 24–29 (2016)
3. Liu, D.: Prediction and analysis of forest fire based on machine learning. Stat. Appl. **5**(2), 163–171 (2016)
4. Cai, X.L., Zhang, G.: Decision scheme of forest fire command and rescue based on arcinfo technology. Hum. For. Sci. Technol. **32**(2), 17–19 (2005)
5. Li, X.: Selection of LS-LVM model and its application in forest fire prediction. Stat. Decis. Making, 163–164 (2011)
6. Wang, Y.M., Yang, J.B., Xu, D.L.: Consumer preference prediction by using a hybrid evidential reasoning and belief rule-based methodology. Expert Syst. Appl. **36**(4), 8421–8430 (2009)
7. Xu, D.L., Liu, J., Yang, J.B.: Inference and learning methodology of belief-rule-based expert system for pipeline leak detection. Expert Syst. Appl. **32**(1), 103–113 (2007)
8. Kong, G., Xu, D.L., Body, R.: A belief rule-based decision support system for clinical risk assessment of cardiac chest pain. Eur. J. Oper. Res. **219**, 564–573 (2012)
9. Chutia, R., Datta, D.: Probability-credibility health risk assessment under uncertain environment. Stoch. Environ. Res. Risk Assess. **31**(2), 449–460 (2017)

10. Dong, W.Y., Li, Q.Y., Yu, H.Y.: Implementation of uncertain reasoning algorithm based on Subjective Bayes method. Technol. Inf. **27**(12), 196–197 (2014)
11. Zhang, Y.J., Wang, X.X., Wang, Y.: A forecasting method of spare parts demand in wartime based on evidence theory. J. Syst. Simul. 1–9 (2018)
12. Zhou, Z.J., Chen, Y.W., Hu, C.H.: Evidence Reasoning, Confidence Rule Base and Complex System Modeling, 1st edn. Science Press, Beijing (2017)
13. Wang, Y.M., Yang, J.B., Xu, D.L.: Environmental impact assessment using the evidential reasoning approach. Eur. J. Oper. Res. **174**(3), 1885–1913 (2006)
14. Zhang, K.: Emergency scheme generation method based on belief rule base. J. Fujian Inst. Eng. **13**(6), 584–589 (2015)
15. Yang, J.B., Liu, J., Xu, D.L.: Optimization models for training belief-rule- based systems. IEEE Trans. Syst. Man Cybern. Part A Syst. Hum. **37**(4), 569–585 (2007)
16. Wang, Y.M., Yang, J.B., Xu, D.L.: Consumer preference prediction by using a hybrid evidential reasoning and belief rule-based methodology. Expert Systems with Applications **36**(4), 8421–8430 (2009)

Kansei Knowledge Extraction as Measure of Structural Heterogeneity

Mina Ryoke[✉] and Tadahiko Sato

Faculty of Business Sciences, University of Tsukuba, Tokyo, Japan
ryoke.mina.ge@u.tsukuba.ac.jp, sato@gssm.otsuka.tsukuba.ac.jp

Abstract. Representative measurements way of affective attributes is the Semantic Differential (SD) method in Kansei evaluation experiments. Structural heterogeneity of the subjective evaluations indicates the heterogeneous evaluation structure of each evaluator, which is presented individually by the specific selected factors. The objective of the structural heterogeneity modeling is not only to extract the general trends but also to identify the diverse individual evaluation structure. In this paper, we propose a Hierarchical Bayes Regression model with Heterogeneous Variable Selection (HBRwHVS), which simultaneously analyses the individual models and identifies the influential explanatory variables based on the framework of the Hierarchical Bayes Regression Modeling. The results offer the relational data between the evaluators and the selected items as carefully chosen explanatory variables. We apply the proposed method to the analysis of sensibility subjective evaluation data on traditional craft and show its effectiveness. After obtaining the estimated values of model parameters, cluster analysis is performed on subjects with the similar evaluation structures as another example of its applications.

Keywords: Semantic Differential method · Structural heterogeneity
HBRwHVS · Clustering · Traditional craft

1 Introduction

It is becoming more convenient to browse modest impressions and full-fledged reviews by reviewers of products on the Internet. The information thus obtained would benefit by adopting "velocity" of the three V's (velocity, volume, and variety) that are associated with big data [20]. However, in situations in which customer preference varied, we should focus more on customer behavior. For example, there exists the structural heterogeneity of evaluation. To obtain knowledge regarding the evaluation structures for products of high concern, importance of the evaluation experiments still alive and the data from heterogeneous subjective evaluations shall be analyzed.

Kansei Engineering [12] analyses customer perceptions and product designs, where customer satisfaction and purchase intentions are influenced not only by

© Springer Nature Singapore Pte Ltd. 2018
J. Chen et al. (Eds.): KSS 2018, CCIS 949, pp. 142–157, 2018.
https://doi.org/10.1007/978-981-13-3149-7_11

the quality but also the consumer preferences. Comprehensive evaluations of customer against products given the various situations are diverse, and many researches on the construction of evaluation models are being conducted in the various fields of Kansei engineering, ergonomics, the sensory evaluation experiment. One result of subjective evaluation data processing at Kansei Engineering is the identification of the relationship between product overall evaluation and sensitivity evaluation words [8,16,18]. Various modelling methods are being studied about expressing individual differences and linking them to product development that meets customer needs [10,14]. Many works contribute to treatment of fuzziness in evaluation [1,3,9,19]. However, this paper focuses on the heterogeneity of evaluators that possess the different linguistic criteria in the evaluation of products.

In this paper, we propose a Hierarchical Bayes Regression model with Heterogeneous Variable Selection (HBRwHVS), which analyses individual models identifying the explanatory variables simultaneously based on the framework of the Hierarchical Bayes Regression modeling. The result of this method offers the relation data between the evaluators and the selected items as carefully chosen explanatory variables. After obtaining the estimated values of the model parameters, cluster analysis is performed on subjects having similar evaluation structure as one of effective utilization examples. We apply the proposed method to the analysis of sensibility subjective evaluation data on traditional craft and show its effectiveness. As a result, it is possible to obtain the priority of the sensitivity evaluation items of people. Here, the aim of this paper is to obtain more valuable structure of each person more aggressively rather than obtaining an evaluation structure by just handling all the evaluation items given equally.

This article consists of the following sections. After introducing the related literature in Sect. 2, data and basic structure of the model are described in Sect. 3. In Sect. 4, we propose HBRwHVS and its estimation algorithm for our model. Section 5 demonstrates the application for traditional craft products, and Sect. 6 describes the conclusion and future work of our research.

2 Literature

In the field of marketing, enormous researches have been carried out regarding the framework of a hierarchical Bayesian model. Many studies aim to assess the consumer heterogeneity, which plays a central role in field of marketing research and practice. The 4P's (Product, Price, Promotion, and Place) tactic of the marketing field, for example, is built upon extracting the differences between consumers with respect to price expectations, promotion sensitivity, valuation of product attributes, etc. In addition, micro-marketing activities towards specific individuals require parameter estimates at the individual level. Therefore, the modeling of consumer heterogeneity has become a central concept in the current on marketing science research (cf. [15]). However, few studies have focused on the evaluation of a heterogeneous variable selection of consumers, which is a remaining problem in this area.

In [5,6], a stochastic search variable selection (SSVS) procedure is proposed for aggregate linear models. This research is a pioneering study of variable selection in Bayesian modeling, but it does not address heterogeneous variable selection. The most commonly utilized sparse modeling in statistics is the Least Absolute Shrinkage and Selection Operator (LASSO) proposed by [17]. LASSO is a method of estimating parameters by maximizing a regularized log-likelihood function obtained by adding the L_1 norm of a parameter to a log-likelihood function. By adopting this approach, it is possible to select variables of higher dimensions. However, this approach, similar to that in [5,6], does not address heterogeneous variable selection.

In [7], the study extends these researches to heterogeneous variable selection. The authors [7] apply the methods to a discrete choice-type conjoint analysis in which data are collected through questionnaire survey. In their model, however, the hierarchical model is not structured and the commonality between consumers behind heterogeneous parameters cannot be evaluated.

3   Preparation

Let us first introduce the data obtained in a subjective evaluation experiment on evaluation targets. The measurement scale is that of Semantic Differential (SD) method [8,13]. The SD method is a measurement method developed by [13] utilized to determine the general semantic dimensions of the event and is often utilized for psychological experiments. The SD scale for this study was implemented as a seven-level bipolar scale.

Question Item x_{jic} : 'left adjective' □□□□□□□ 'right adjective'

The obtained data is the three way data since plural evaluation targets such as traditional crafted cups are evaluated through carefully chosen pairs of adjectives by lots of respondents. The data notation is described as follows:

$$x_{jic} = \{1,2,3,4,5,6,7\}, \quad y_{ic}^{(m)} = \{1,2,3,4,5,6,7\}$$
$$j = 1,2,\cdots,J, \; i = 1,2,...,I, \; c = 1,2,\cdots,C, \; m = 1,2,\cdots,M \quad (1)$$

where I denotes the total number of evaluators, C is the total number of evaluation targets, and y_{ic} is the entity of an $I \times C$ matrix of the responses. As it is assumed that there are various types of purchasing intentions, M denotes the number of total evaluations in terms with various aspects. When M is greater than 1, the individual model of each m-th aspect shall be developed. The symbol x_{jic} represents the regressors which consist of the answer of i-th person in response to the j-th question item for the c-cup. As the demographic data of the respondents (such as the grouped age, the gender, and occupation) vary, the notation $z_{ik}(i = 1,2,\cdots,I, k = 1,2,\cdots,K)$ indicates the dummy variable. When $m = 1$ and the evaluation target is only one product, the response as a total evaluation is described by the following regression model.

$$y = \beta_0 1_n + X\beta + \varepsilon, \qquad (2)$$

where y is the $I \times 1$ vector of responses, β_0 is the intercept, and $\beta = (\beta_1, \cdots, \beta_J)^T$ are the regression parameters. The symbol X represents the $I \times J$ matrix of regressors, and ε is the $I \times 1$ vectors of independent and identically distributed normal errors with the mean 0 and the unknown variance σ^2. It is strongly assumed that all evaluators have the same evaluation structure, and answer all the question items. Next, the extraction of the heterogeneous evaluation structures of individuals based on the three way data, is discussed.

4   Model

In this section, we explain two types of models: Hierarchical Bayes Regression model (HBR) and the Hierarchical Bayes Regression model with Heterogeneous Variable Selection (HBRwHVS). In this study, HBRwHVS is the model proposed on the basis of HBR. HBR is shown in Sect. 4.1, and HBRwHSV is proposed in Sect. 4.2.

4.1   Basic Model

HBR is the base model of our research and consists of three sub models: the observation model (within the individual models), hierarchical model (between the individual models) and the prior distribution of parameters. In the following, let i $(i = 1, \cdots, I)$ and c $(c = 1, \cdots, C)$ be subscripts indicating individuals and products, respectively.

Observation Model

$$y_{ic} = \beta_{0i} + \beta_{1i}x_{1ic} + \cdots + \beta_{Ji}x_{Jic} + \varepsilon_{ic}, \varepsilon_{ic} \sim \mathrm{N}(0, \sigma_i^2) \qquad (3)$$
$$= \boldsymbol{x}_{ic}^t \boldsymbol{\beta}_i + \varepsilon_{ic}$$

where, N presents a normal distribution.

Hierarchical Model

$$\boldsymbol{\beta}_i = \boldsymbol{\theta} \boldsymbol{z}_i + \boldsymbol{v}_i, \boldsymbol{v}_i \sim \mathrm{MVN}(\boldsymbol{0}, \boldsymbol{V}_\beta), \qquad (4)$$

where, MVN is the abbreviation for multivariate normal distribution. Now let q represent the number of regressors for the hierarchical model of β_i. Then $\boldsymbol{z}_i$, $\boldsymbol{\theta}$ and $\boldsymbol{v}_i$ denote a regressor vector $(q \times 1)$, coefficients matrix $((J+1) \times q)$ and error term vector $((J+1) \times 1)$ with mean vector $\boldsymbol{0}$ and $((J+1) \times 1)$ variance-covariance matrix $\boldsymbol{V}_\beta$.

Prior Distributions

$$\sigma_i^2 \sim \mathrm{IG}\left(\frac{a_0}{2}, \frac{b_0}{2}\right), \qquad (5)$$

where IG denotes the inverted Gamma function. The value of parameter a_0 is given as the design parameter and the value of parameter b_0 is dependent on a_0 and the variance of y_i. They are described later in this paper.

$$vec\,(\boldsymbol{\theta})\,|\boldsymbol{V}_\beta \sim \mathrm{MVN}\left(\boldsymbol{0}, \boldsymbol{V}_\beta \otimes \boldsymbol{Q}_\beta^{-1}\right) \qquad (6)$$

The notation $vec()$ function combines each column of the matrix by rows. Then, there is:

$$V_\beta \sim \text{IW}(g_\beta, H_\beta) \tag{7}$$

where, IW is the inverted Wishart distribution. It is a probability distribution defined by real-valued positive-definite matrices. In Bayesian statistics it is used as the conjugate prior to the covariance matrix of a MVN. The parameter values are set as follows:

$$a_0 = 100, b_0 = 100, \boldsymbol{Q}_\beta = 0.01\boldsymbol{E}, g_\beta = 32, \text{H}_\beta = g_\beta \times 0.01 \times \boldsymbol{E} \tag{8}$$

where, $\boldsymbol{E}$ denotes an identity matrix.

The model formulated by Eqs. (3)–(8) can be estimated by using Gibbs sampler [4]. Gibbs sampler is one of the Markov Chain Monte Carlo algorithm (MCMC) method that generates a sample sequence that approximates it instead of the probability distribution that is difficult to directly sample. The generated samples are used to approximate the integral calculations such as joint distribution, marginal distribution, expectation value and so on. Samples generated by the Gibbs sampler are always adopted, therefore, more efficient than general MCMC algorithms such as the Metropolis-Hastings (M-H) method.

In the following steps, let $\boldsymbol{y}_i = (y_{i1}, \cdots, y_{iC})^t$, $\boldsymbol{x}_i = (\boldsymbol{x}_{i1}^t|\boldsymbol{x}_{i2}^t|\cdots|\boldsymbol{x}_{iC}^t)$ and $\boldsymbol{Z} = (\boldsymbol{z}_1, \cdots, \boldsymbol{z}_q)^t$. The symbol "|" means to combine vectors vertically. Gibbs sampler for HBR can be implemented in the following four steps. In the first two steps, heterogeneous parameters $(\boldsymbol{\beta}_i, \sigma_i)$ are drawn and hierarchical parameters are generated in the remaining steps. Estimation of the model repeats this algorithm and obtains joint posterior distribution of parameters based on the generated samples.

Estimation Algorithm for HBR

1. $\boldsymbol{\beta}_i | \sigma_i^2, \boldsymbol{\theta}, \boldsymbol{V}_\beta, y_i, \boldsymbol{x}_i \sim \text{MVN}\left(\tilde{\boldsymbol{\beta}}_i, \sigma_i^2\left(\boldsymbol{x}_i^t \boldsymbol{x}_i + \boldsymbol{V}_\beta^{-1}\right)^{-1}\right)$,

$\tilde{\boldsymbol{\beta}}_i = \left(\boldsymbol{x}_i^t \boldsymbol{x}_i + \boldsymbol{V}_\beta^{-1}\right)^{-1}\left(\boldsymbol{x}_i^t \boldsymbol{x}_i \widehat{\boldsymbol{\beta}}_i + \bar{\boldsymbol{\beta}}_i\right), \bar{\boldsymbol{\beta}}_i = \boldsymbol{\theta}^t \boldsymbol{z}_i, \widehat{\boldsymbol{\beta}}_i = (\boldsymbol{x}_i^t \boldsymbol{x}_i)^{-1}\boldsymbol{x}_i^t y_i$

2. $\sigma_i^2 | \boldsymbol{\beta}_i, \boldsymbol{\theta}, \boldsymbol{V}_\beta, y_i \sim \text{IG}\left(\dfrac{r_i^*}{2}, \dfrac{s_i^*}{2}\right)$

$r_i^* = a_0 + I, s_i^* = b_0 + (y_i - \boldsymbol{x}_i^t \boldsymbol{\beta}_i)^t (y_i - \boldsymbol{x}_i^t \boldsymbol{\beta}_i)$

3. $\boldsymbol{\delta} | \boldsymbol{\beta}_i, \boldsymbol{V}_\beta, \boldsymbol{Z} = \text{vec}(\boldsymbol{\delta}) | \boldsymbol{\beta}_i, \boldsymbol{V}_\beta, \boldsymbol{Z} \sim \text{MVN}\left(\tilde{\boldsymbol{d}}, \boldsymbol{V}_\beta \otimes (\boldsymbol{Z}^t \boldsymbol{Z} + \boldsymbol{Q}_\beta)^{-1}\right)$

$\tilde{\boldsymbol{d}} = \text{vec}\left(\tilde{\boldsymbol{D}}\right), \tilde{\boldsymbol{D}} = (\boldsymbol{Z}^t \boldsymbol{Z} + \boldsymbol{Q}_\beta)^{-1}\left(\boldsymbol{Z}^t \boldsymbol{Z} \widehat{\boldsymbol{D}} + \boldsymbol{Q}_\beta \bar{\boldsymbol{D}}\right)$,

$\widehat{\boldsymbol{D}} = (\boldsymbol{Z}^t \boldsymbol{Z})^{-1} \boldsymbol{Z}^t \boldsymbol{B}, \bar{\boldsymbol{D}} = \text{vec}\left(\widehat{\boldsymbol{\delta}}\right)$

4. $\boldsymbol{V}_\beta | \boldsymbol{\beta}_i, \sigma_i^2, \boldsymbol{\delta}, y_i \sim \text{IW}(g_\beta + I, H_\beta + \boldsymbol{S})$

$\boldsymbol{S} = \displaystyle\sum_{i=1}^{I}\left(\boldsymbol{\beta}_i - \bar{\boldsymbol{\beta}}_i\right)\left(\boldsymbol{\beta}_i - \bar{\boldsymbol{\beta}}_i\right)^t, \bar{\boldsymbol{\beta}}_i = \boldsymbol{\delta}^t \boldsymbol{z}_i$

4.2   Proposed Model

We propose a model to evaluate the consumers' heterogeneous structure of product evaluation. From statistical point of view, the extraction of the consumers' heterogeneous structure corresponds to the heterogeneous variable selection. The proposed model can be formulated by extending Eq. (3) to (8).

Observation Model

$$y_{ic} = \beta_{0i}\tau_{0i} + \beta_{1i}\tau_{1i}x_{1ic} + \cdots + \beta_{Ji}\tau_{Ji}x_{Jic} + \varepsilon_{ic}, \varepsilon_{ic} \sim N(0, \sigma_i^2) \qquad (9)$$
$$= x_{ic}^t T_i \beta_i + \varepsilon_{ic}$$

where, T_i is a $(J+1) \times (J+1)$ matrix, formed as $\mathrm{diag}(\tau_{0i}, \cdots, \tau_{pi})$ and $\tau_{ji} \in \{\xi, 1\}$. The symbol ξ is a constant that is small enough such that $T_i \theta z_i$ is very close to zero when $\tau_{ji} = \xi$. This assumption is equivalent to assuming the hierarchical model in Eq. (10).

$$\beta_i \sim N(T_i \theta z_i, T_i V_\beta T_i) \qquad (10)$$

The advantage of this specification is that it leads to standard Bayesian updating formulas conditional on τ because $T_i^{-1}\beta_i \sim N(\theta z_i, V_\beta)$. That is, $\beta_i^* = T_i^{-1}\beta_i$ is used to draw θ and V_β. Eq. (11) becomes a hierarchical model of β_i. Heterogeneity in τ_{ji} is accommodated by assuming $\tau_{ji} = 1$ with probability p_j and it equals ξ with probability $1 - p_j$, with prior probability Beta$(a_{\tau,0}, b_{\tau,0})$, shown by Eq. (17).

Hierarchical Model

$$\beta_i^* \sim N(\theta z_i, V_\beta) \qquad (11)$$

Prior Distributions

$$\sigma_i^2 \sim IG\left(\frac{a_0}{2}, \frac{b_0}{2}\right) \qquad (12)$$

$$vec\,(\theta)\,|V_\beta \sim \mathrm{MVN}\left(0, V_\beta \otimes Q_\beta^{-1}\right) \qquad (13)$$

$$V_\beta \sim \mathrm{IW}\,(g_\beta, H_\beta) \qquad (14)$$

$$p_j \sim \mathrm{Beta}\,(a_{\tau_{ji},0}, b_{\tau_{ji},0}) \qquad (15)$$

where, Beta denotes a beta distribution.

$$a_{\tau_{ji},0} = 8, b_{\tau_{ji},0} = 4 \qquad (16)$$

The estimation of this model cannot apply to the Gibbs sampler in Sect. 4.1, because of not being conjugacy. The joint posterior distribution should be decomposed to derive an algorithm of the model. We can obtain the algorithm for the model based on Eq. (17). These are assumptions for modeling we propose.

$$
\begin{aligned}
& f\left(\{\boldsymbol{\beta}_i\}, \{\boldsymbol{\tau}_i\}, \{\sigma_i^2\}, \boldsymbol{\theta}, \boldsymbol{V}_\beta, \boldsymbol{p} \mid \{\boldsymbol{y}_{ic}\}, \{\boldsymbol{x}_i\}, \{\boldsymbol{z}_i\}\right) \\
& \propto f(\boldsymbol{p}) f(\boldsymbol{\theta}|\boldsymbol{V}_\beta) f(\boldsymbol{V}_\beta) \prod_{i=1}^{I} f\left(\boldsymbol{\beta}_i|\boldsymbol{\tau}_i, \boldsymbol{\theta}, \boldsymbol{V}_\beta, \boldsymbol{z}_i, \sigma_i^2\right) f(\boldsymbol{\tau}_i|\boldsymbol{p}) f\left(\sigma_i^2\right) \times \\
& \qquad\qquad \prod_{c=1}^{C} f\left(y_{ic}|\boldsymbol{\beta}_i, \boldsymbol{\tau}_i, \boldsymbol{x}_i, \sigma_i^2\right) \\
& = \underbrace{\prod_{j=0}^{J} f(p_j)}_{\text{Beta}} \underbrace{f(\boldsymbol{\theta}|\boldsymbol{V}_\beta)}_{\text{MVN}} \underbrace{f(\boldsymbol{V}_\beta)}_{\text{IW}} \underbrace{\prod_{i=1}^{I} f\left(\boldsymbol{\beta}_i^*|\boldsymbol{\tau}_i, \boldsymbol{\theta}, \boldsymbol{V}_\beta, \boldsymbol{z}_i\right)}_{\text{MVN}} \underbrace{\prod_{j=0}^{J} f(\tau_{ji}|p_j)}_{\text{Bernoulli}} \underbrace{f\left(\sigma_i^2\right)}_{\text{IG}} \times \\
& \qquad\qquad \underbrace{\prod_{c=1}^{C} f\left(y_{ic}|\boldsymbol{\beta}_i, \boldsymbol{\tau}_i, \boldsymbol{x}_i, \sigma_i^2\right)}_{\text{Normal}}
\end{aligned}
\tag{17}
$$

The algorithm for HBRwHVS can be implemented in the following five steps. In the first two steps, heterogeneous parameters $(\boldsymbol{\beta}_i, \boldsymbol{\tau}_i, \sigma_i)$ are drawn and hierarchical parameters are generated in the remaining steps. In first step, we generate $\boldsymbol{\beta}_i, \boldsymbol{\tau}_i$ by using an independent M-H sampler. This algorithm consists of three steps. Step 2 $\sim$ Step 4 are the same algorithm for HBR. The final step can be derived from the conjugacy relationship between the Bernoulli (Binomial) likelihood and beta prior. Estimation of the model repeats this algorithm and obtains joint posterior distribution of parameters based on the generated samples. We call this algorithm M-H within Gibbs the sampler (hybrid sampler).

Estimation Algorithm for HBRwHVS

1. Generate $\boldsymbol{\beta}_i, \boldsymbol{\tau}_i|\sigma_i^2, \boldsymbol{\theta}, \boldsymbol{V}_\beta, y_i, \boldsymbol{x}_i$ by using independent M-H sampler
   (a) $\tau_{ji}^\dagger \sim \text{Ber}(p_j^{(r-1)}), j = 0, \cdots, J$. Ber indicates Bernoulli distribution with probability p_j.
   (b) $\boldsymbol{\beta}_i^\dagger \sim \text{N}(\boldsymbol{T}_i^\dagger \boldsymbol{\theta}^{(r-1)} \boldsymbol{z}_i, \boldsymbol{T}_i^\dagger \boldsymbol{V}_\beta^{(r-1)} \boldsymbol{T}_i^\dagger)$
   (c) Accept the new values $\tau_{ji}^{(r)}$ and $\boldsymbol{\beta}_i^{(r)}$ with the probability
   $$
   \text{Pr(accept)} = \min\left(\frac{\text{L}_i(\tau_{ji}^{(\dagger)}, \boldsymbol{\beta}_i^{(\dagger)})}{\text{L}_i(\tau_{ji}^{(r-1)}, \boldsymbol{\beta}_i^{(r-1)})}, 1\right).
   $$
   $\text{L}_i(\cdot)$ is the likelihood of the data for individual i.

2. $\sigma_i^2|\boldsymbol{\beta}_i, \boldsymbol{\theta}, \boldsymbol{V}_\beta, y_i \sim \text{IG}\left(\frac{r_i^*}{2}, \frac{s_i^*}{2}\right)$
   $r_i^* = a_0 + I, s_i^* = b_0 + (y_i - \boldsymbol{x}_{ic}^t \boldsymbol{T}_i \boldsymbol{\beta}_i)^t (y_i - \boldsymbol{x}_{ic}^t \boldsymbol{T}_i \boldsymbol{\beta}_i)$

3. $\text{vec}(\boldsymbol{\delta})|\boldsymbol{\beta}_i^{(*),(r)}, \boldsymbol{V}_\beta, \boldsymbol{Z} \sim \text{MVN}\left(\tilde{\boldsymbol{d}}, \boldsymbol{V}_\beta \otimes (\boldsymbol{Z}^t \boldsymbol{Z} + \boldsymbol{Q}_\beta)^{-1}\right)$
   $\tilde{\boldsymbol{d}} = \text{vec}\left(\tilde{\boldsymbol{D}}\right), \tilde{\boldsymbol{D}} = (\boldsymbol{Z}^t \boldsymbol{Z} + \boldsymbol{Q}_\beta)^{-1}\left(\boldsymbol{Z}^t \boldsymbol{Z} \hat{\boldsymbol{D}} + \boldsymbol{Q}_\beta \bar{\boldsymbol{D}}\right),$
   $\hat{\boldsymbol{D}} = (\boldsymbol{Z}^t \boldsymbol{Z})^{-1} \boldsymbol{Z}^t \boldsymbol{B}^{(*),(r)}, \bar{\boldsymbol{D}} = \text{vec}\left(\hat{\boldsymbol{\delta}}\right)$

4. $V_\beta | \beta_i^{(*),(r)}, \sigma_i^2, \delta, y_i \sim \text{IW}(g_\beta + I, H_\beta + S)$

$$S = \sum_{i=1}^{I} \left(\beta_i^{(*),(r)} - \bar{\beta}_i \right) \left(\beta_i^{(*),(r)} - \bar{\beta}_i \right)^t, \bar{\beta}_i = \delta^t z_i$$

5. Generate $p_j | \{\tau_{ji}\}$ for $j = 1, \cdots, J$.Let $s_{ji} = 1$ if $\tau_{ji} = 1$ and 0 if otherwise.

$$p_j \sim \text{Beta}(a_{\tau,0} + \sum_{i=1}^{I} s_{ji}, I - \sum_{i=1}^{I} s_{ji} + b_{\tau,0})$$

5   Application in Traditional Craft Product

5.1   Data

Traditional crafted coffee cups (examples are shown in Fig. 1) are the evaluation targets, here. The number of prepared cups is 40 ($C = 40$). Respondents are asked to provide their answers regarding each evaluation item. The number of the respondents is 77 ($I = 77$). The Kansei evaluation data is obtained by a subjective evaluation experiment.

Fig. 1. Examples of evaluation targets

The SD method is a measurement method for measuring the general semantic dimension of the event developed by [13] and is often used for psychological experiments. The SD data for this study was acquired on a 7-level bipolar scale. At the same time, respondent demographic data such as gender, age, occupation, experience of purchase and so forth are also collected. Thus $z_{ik}(k = 1, \cdots, q)$ show 1 when a respondent has the attribute, 0 otherwise.

We have three types of dependent variables $y_{ic} = \{1, 2, 3, 4, 5, 6, 7\}$, for traditional craft evaluations: $y_{ic}^{(1)}$ (happy if given this as a gift), $y_{ic}^{(2)}$ (want to buy it for myself) and $y_{ic}^{(3)}$ (want to buy it for others). The evaluation of these cups, in general, highly depends on the individual's preference. We model the relationship between each consumer's intention and the explanatory variables and analyze the difference in mechanisms for between individuals. There are three types of explanatory variables used in each model: Kansei (15 variables, SD), design impressions (5 variables, SD), and product characteristics (8 variables, dummy variables). The regressors x_{jic} include not only the subjective evaluation data but also the dummy variables, which express the hard features of each cup, i.e., $x_{jic} = \{1, 2, 3, 4, 5, 6, 7\}$, $(j = 1, \cdots, 20)$, $x_{jic}(j = 21, \cdots, 28)$ show the dummy variables. The questionnaire items in the answering sheets have been designed taking into account customers voice in retail shops in daily. Table 1 shows a summary of the variables. By applying the proposed method, we can get the estimators of the regressors $\beta_i^{(m)}$ ($i = 1, \cdots, I$, $m = 1, 2, 3$). There are rich variations on how to analyze the obtained estimators, as there are the total

Table 1. Variables

Variable	Type	Group	Description	Supplement
$y_{ic}^{(1)}$	Ordinary	Intention	happy if given this as a gift	
$y_{ic}^{(2)}$	Ordinary	Intention	want to buy it for myself	
$y_{ic}^{(3)}$	Ordinary	Intention	want to buy it for others	
x_{1ic}	SD	Kansei	soft$\Longleftrightarrow$hard	
x_{2ic}	SD	Kansei	warm-heated$\Longleftrightarrow$refreshing	
x_{3ic}	SD	Kansei	lively$\Longleftrightarrow$quiet	
x_{4ic}	SD	Kansei	adorable$\Longleftrightarrow$sober	
x_{5ic}	SD	Kansei	neat$\Longleftrightarrow$unique	
x_{6ic}	SD	Kansei	sunny$\Longleftrightarrow$subdued	
x_{7ic}	SD	Kansei	bright$\Longleftrightarrow$calm	
x_{8ic}	SD	Kansei	deluxe$\Longleftrightarrow$simple	
x_{9ic}	SD	Kansei	friendly$\Longleftrightarrow$ceremonious	
x_{10ic}	SD	Kansei	modern$\Longleftrightarrow$classic	
x_{11ic}	SD	Kansei	cheerful$\Longleftrightarrow$dignified	
x_{12ic}	SD	Kansei	never get tired$\Longleftrightarrow$trend	
x_{13ic}	SD	Kansei	gently$\Longleftrightarrow$bracing	
x_{14ic}	SD	Kansei	delicate$\Longleftrightarrow$relaxed	
x_{15ic}	SD	Kansei	relief$\Longleftrightarrow$strained	
x_{16ic}	SD	Design	lightweight$\Longleftrightarrow$heavyweight	
x_{17ic}	SD	Design	plump $\Longleftrightarrow$slender	
x_{18ic}	SD	Design	thin$\Longleftrightarrow$thick	
x_{19ic}	SD	Design	rounded design$\Longleftrightarrow$liner design	
x_{20ic}	SD	Design	smaller$\Longleftrightarrow$largish	
x_{21ic}	Dummy	Product	traditional pattern	Common variables
x_{22ic}	Dummy	Product	demitasse	Common variables
x_{23ic}	Dummy	Product	earthware texture	Common variables
x_{24ic}	Dummy	Product	whitebase	Common variables
x_{25ic}	Dummy	Product	patternbase	Common variables
x_{26ic}	Dummy	Product	five colors of kutaniware	Common variables
x_{27ic}	Dummy	Product	blue and white porcelain	Common variables
x_{28ic}	Dummy	Product	aka-e	Common variables

evaluations in terms of the three aspects of the products. We can analyze the estimators with respect to each response when one of them is the main concern. This paper focuses on the individual evaluation structure and we get started by showing the counts of the selected items by the evaluators in Table 2. By combining the chunk of estimators on each response by rows, the differences or common of the contribution to each response variable are derived through the regressors.

5.2 Results of the Estimation

The proposed method can provide lots of information such as the selected variables of each individual and the estimators of each personalized model. In Table 2, the numbers of the variables utilized by each individual are shown. The numbers are maximum and minimum numbers of the variables used. The maximum number is 29, which includes the constant term and all variables. The variety in numbers is observed. Some evaluators consider fewer variables than we prepared before the subjective evaluation experiment.

Table 2. Counts of selected items by each respondant

Individ.	1	2	3	4	5	6	7	8	9	10	11	12	13	14	15	16	17	18	19	20	21	22	23	24	25	26
$y_{ic}^{(1)}$	21	22	21	26	26	25	20	24	23	18	27	26	21	23	24	23	18	28	20	21	29	26	25	22	27	28
$y_{ic}^{(2)}$	24	22	24	23	24	23	18	22	26	22	20	22	22	22	25	24	21	22	24	23	27	17	24	21	19	26
$y_{ic}^{(3)}$	27	29	25	25	27	16	22	24	25	25	25	23	22	27	26	28	25	20	23	23	27	25	26	22	20	28

Individ.	27	28	29	30	31	32	33	34	35	36	37	38	39	40	41	42	43	44	45	46	47	48	49	50	51	52
$y_{ic}^{(1)}$	23	25	26	27	27	22	23	25	24	24	21	22	24	24	20	27	25	24	23	27	26	25	18	24	20	21
$y_{ic}^{(2)}$	22	26	23	23	28	23	25	24	21	23	26	27	25	22	22	26	22	24	22	24	25	24	18	26	21	22
$y_{ic}^{(3)}$	19	28	27	16	27	25	27	19	21	22	24	22	20	26	19	28	22	27	19	28	22	25	27	25	22	24

Individ.	53	54	55	56	57	58	59	60	61	62	63	64	65	66	67	68	69	70	71	72	73	74	75	76	77
$y_{ic}^{(1)}$	27	20	22	22	24	25	21	20	24	25	26	21	29	24	26	24	22	27	26	27	26	26	22	25	20
$y_{ic}^{(2)}$	22	23	24	19	23	23	20	24	18	25	22	24	25	25	25	22	27	23	23	27	23	22	25	26	17
$y_{ic}^{(3)}$	18	24	25	25	27	26	23	25	22	28	18	27	23	23	20	28	23	27	25	27	27	22	28	25	23

Variable Selection. This section discusses the estimation results. By estimating the model, we can distinguish between "positively influencing variables", "negatively influencing variables" and "non-influencing variables". Table 3 is the percentage of individuals in the above three categories for each dependent variable. Gray shading indicates cells with a ratio greater than 50%. There are many individuals whose adjective pair "adorable⟺sober " positively influences the dependent variables, and the ratio is not different between the three dependent variables. On the other hand, the adjective pair "neat ⟺ unique" differs from the former case. As the evaluation of an adjective for the traditional craft approaches "unique", $y_{ic}^{(1)}$ (happy if given a gift) and $y_{ic}^{(2)}$ (want to buy it for myself) tend to increase. However, as the evaluation of adjective for the traditional craft approaches "neat", $y_{ic}^{(3)}$ (want to buy it for others) tends to increase. Other variables can also be verified by the same method, but are omitted for simplicity and space constraint.

Figure 2 shows the scatter plots of the responses $(\beta_{5i} \times \tau_{5i})$ for the adjective pair "neat⟺unique" against three dependent variables $(y_{ic}^{(1)}, y_{ic}^{(2)}, y_{ic}^{(3)})$. The left, central and right panels correspond to $(\beta_{5i}^{(2)} \times \tau_{5i}^{(2)})$ against $(\beta_{5i}^{(1)} \times \tau_{5i}^{(1)})$, $(\beta_{5i}^{(3)} \times \tau_{5i}^{(3)})$ against $(\beta_{5i}^{(1)} \times \tau_{5i}^{(1)})$ and $(\beta_{5i}^{(3)} \times \tau_{5i}^{(3)})$ against $(\beta_{5i}^{(2)} \times \tau_{5i}^{(2)})$, respectively. In addition, a superscript $(j), j = 1, 2, 3$ indicates the parameters of the model that corresponds to $y_{ic}^{(1)}$, $y_{ic}^{(2)}$ or $y_{ic}^{(3)}$. Each point in the panels corresponds to an individual. The points concentrate in the first and the third quadrant, and the three panels suggest a positive correlation. However, as shown in the central and right panels, there are many points in the second quadrant. This indicates that even for the same explanatory variable, there are individuals that have the opposite response for different objective variables. Figure 3 depicts the scatter plots of the responses $(\beta_{10i} \times \tau_{10i})$ for the adjective pair "modern⟺classic", and the panel can be viewed in the same way as Fig. 2. The responses to this adjective pair is concentrated in the first quadrant, which is different from that of Fig. 2.

Table 3. Ratio

explanatory variables	happy if given this as a gift			want to buy it for myself			want to buy it for others		
	+	−	0	+	−	0	+	−	0
Const.	0.7922	0.2078	0.0000	0.2208	0.7792	0.0000	0.2597	0.7403	0.0000
soft⟺hard	0.6364	0.1558	0.2078	0.4026	0.2338	0.3636	0.6234	0.1558	0.2208
warm-heated⟺refreshing	0.3636	0.2338	0.4026	0.3117	0.2468	0.4416	0.2078	0.2078	0.5844
lively⟺quiet	0.0909	0.8442	0.0649	0.2857	0.5844	0.1299	0.2208	0.6883	0.0909
adorable⟺sober	0.9091	0.0519	0.0390	0.9221	0.0649	0.0130	0.9221	0.0649	0.0130
neat⟺unique	0.3377	0.5325	0.1299	0.3377	0.4416	0.2208	0.5974	0.3117	0.0909
brilliant⟺subdued	0.3766	0.2857	0.3377	0.5714	0.3636	0.0649	0.6104	0.1169	0.2727
bright⟺calm	0.3636	0.5455	0.0909	0.1558	0.3247	0.5195	0.4545	0.4286	0.1169
deluxe⟺simple	0.8312	0.1169	0.0519	0.7662	0.1688	0.0649	0.9091	0.0649	0.0260
friendly⟺ceremonious	0.2857	0.4805	0.2338	0.4805	0.3896	0.1299	0.3247	0.5455	0.1299
modern⟺classic	0.8831	0.0649	0.0519	0.8701	0.1169	0.0130	0.7662	0.1039	0.1299
cheerful⟺dignified	0.2208	0.6883	0.0909	0.4156	0.2857	0.2987	0.4675	0.3896	0.1429
never get tired⟺trend	0.9351	0.0519	0.0130	0.9351	0.0260	0.0390	0.9610	0.0000	0.0390
gently⟺bracing	0.4545	0.3377	0.2078	0.1429	0.1169	0.7403	0.5325	0.3636	0.1039
delicate⟺relaxed	0.7532	0.1429	0.1039	0.6364	0.1818	0.1818	0.6623	0.2987	0.0390
relief⟺strained	0.3117	0.6104	0.0779	0.4416	0.2987	0.2597	0.2338	0.4026	0.3636
lightweight⟺heavyweight	0.2727	0.4026	0.3247	0.3247	0.5974	0.0779	0.1948	0.3247	0.4805
plump ⟺slender	0.4805	0.4545	0.0649	0.5714	0.3247	0.1039	0.3766	0.4935	0.1299
thin⟺thick	0.2338	0.7013	0.0649	0.2338	0.6753	0.0909	0.3377	0.5455	0.1169
rounded design⟺liner design	0.5065	0.2597	0.2338	0.6364	0.2597	0.1039	0.5844	0.0519	0.3636
smaller⟺largish	0.1818	0.6104	0.2078	0.1948	0.4416	0.3636	0.3247	0.5455	0.1299
traditional pattern	0.2078	0.2338	0.5584	0.2208	0.7662	0.0130	0.2468	0.4805	0.2727
demitasse	0.4416	0.4026	0.1558	0.2597	0.1688	0.5714	0.4805	0.3117	0.2078
earthware texture	0.1299	0.0909	0.7792	0.6883	0.2208	0.0909	0.4286	0.5195	0.0519
whitebase	0.1818	0.6753	0.1429	0.4416	0.3636	0.1948	0.2857	0.6494	0.0649
patternbase	0.4545	0.3766	0.1688	0.3377	0.1039	0.5584	0.3377	0.4026	0.2597
five colors of kutaniware	0.2597	0.5065	0.2338	0.2468	0.6494	0.1039	0.1948	0.6494	0.1558
blue and white porcelain	0.8182	0.1169	0.0649	0.8182	0.0909	0.0909	0.8052	0.1558	0.0390
aka-e	0.6104	0.2857	0.1039	0.6234	0.3117	0.0649	0.2468	0.4805	0.2727

As the evaluation of adjectives for the traditional craft approaches "modern", consumer intention $(y_{ic}^{(1)}, y_{ic}^{(2)}, y_{ic}^{(3)})$ tend to increase on average. However, there exist consumers that responded to the adjective for $y_{ic}^{(3)}$ with zero (unaffected variable). These are shown on the $x-$axis of the central and right panels.

The above findings are useful for realizing the sophisticated micro-marketing activities of companies or developing traditional crafts that are highly acclaimed by consumers. Micro-marketing is a marketing strategy in which marketing efforts are focused on a small group, specific individuals. These findings cannot be obtained with conventional statistical approaches, but can only be obtained by adopting the proposed model.

Structural Heterogeneity Among Evaluators. Our next concern is grouping which shows similar evaluation structures on the models for purchase intention among the evaluators. Based on the obtained estimators, the binary matrix be derived with 1 as the selected item and 0 as unselected item, and the real-value matrix shows the estimators, i.e., the relation data. Here, we focus on the binary matrix and describe the heatmap. The heatmap provides the results of Ward's

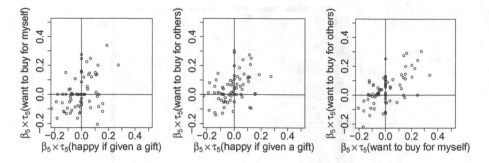

Fig. 2. Posterior mean of β_{5i} for each dependent variables

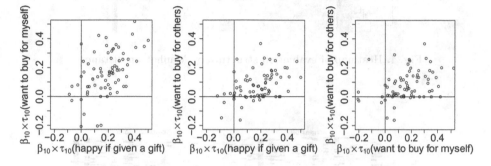

Fig. 3. Posterior mean of β_{10i} for each dependent variables

method simultaneously, and the distance between respondents is determined by the Manhattan distance formula. Figure 4 shows the heatmap that was obtained to visualize the evaluators who have a similar evaluation structure. Obviously light and dark shade is observed.

Simultateous Analysis of Evaluators and Selected Variables by Co-clustering. Next, we analyze the relation data between the evaluation words and the respondents by using the co-clustering method (finite mixture model) [2, 11]. Analysis results are shown in Figs. 5 and 6. Figure 5 illustrates the binary matrix as the relation between the items and the respondents. The white area denotes 1 as the selected item, while black is 0, the unselected item. Here, the number of clusters is given as (2,2) for convenience. Further analyses with different cluster numbers shall be conducted.

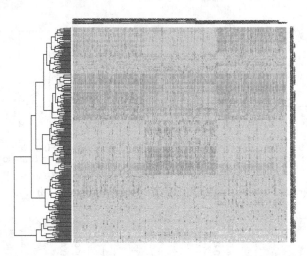

Fig. 4. Heatmap of evaluation structure by Manhattan distance

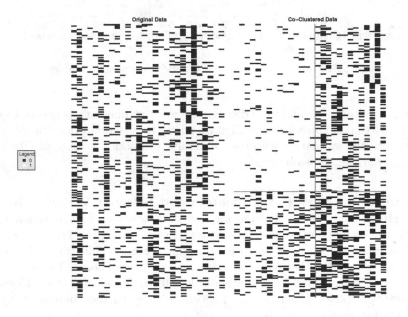

Fig. 5. Original and co-clustered binary data

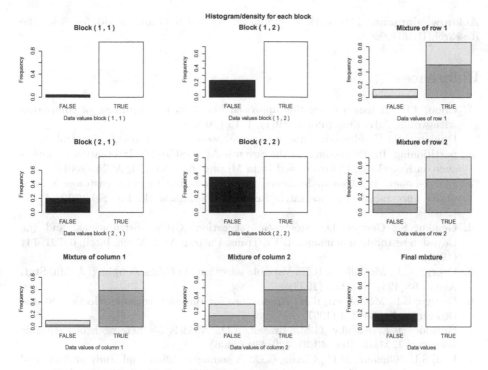

Fig. 6. Distributions for each block along with various mixture densities

6   Conclusion and Future Work

The proposed method has been applied to subjective evaluation data regarding evaluation of traditional crafted cups. The obtained results include variable selections to show the heterogeneous evaluation structure of each individual. After demonstrating the consideration of the obtained estimators, the relation data has been analyzed by heatmap and co-clustering. The existence of several clusters would guide the understanding of the potential customers and development of variety of products. There is structural heterogeneity not only among the individuals but also the response variables, as observed from the selected or unselected regressors in binary terms, or the sign of estimator values with respect to the adjectives scaled by the SD method.

Thus, future works shall involve the identification of the relationship between evaluation targets and evaluation words through the proposed method. The model would suggest the eligible combination of words that should be utilized to express the corresponding evaluation targets. The results would support store owners in recommending products by taking into account the words spoken by customers and assist then in suggesting new product developments by considering the high prioritized words used by their clients.

Acknowledgment. This work is supported by JSPS Grand-in-Aid for Scientific Research(B)18H00904.

References

1. Chou, J.R.: A kansei evaluation approach based on the technique of computing with words. Adv. Eng. Inform. **30**(1), 1–15 (2016)
2. Dhillon, I.S.: Co-clustering documents and words using bipartite spectral graph partitioning. In: Proceedings of the Seventh ACM SIGKDD International Conference on Knowledge Discovery and Data Mining. pp. 269–274. ACM (2001)
3. Florez-Lopez, R., Ramon-Jeronimo, J.M.: Managing logistics customer service under uncertainty: an integrative fuzzy Kano framework. Inf. Sci. **202**, 41–57 (2012)
4. Geman, S., Geman, D.: Stochastic relaxation, Gibbs distributions, and the Bayesian restoration of images. IEEE Trans. Pattern Anal. Mach. Intell. **6**, 721–741 (1984)
5. George, E.I., McCulloch, R.E.: Variable selection via Gibbs sampling. J. Am. Stat. Assoc. **88**(423), 881–889 (1993)
6. George, E.I., McCulloch, R.E.: Approaches for Bayesian variable selection. Statistica Sinica **7**, 339–373 (1997)
7. Gilbride, T.J., Allenby, G.M., Brazell, J.D.: Models for heterogeneous variable selection. J. Mark. Res. **43**(3), 420–430 (2006)
8. Hsu, S.H., Chuang, M.C., Chang, C.C.: A semantic differential study of designers' and users' product form perception. Int. J. Ind. Ergon. **25**(4), 375–391 (2000)
9. Huang, Y., Chen, C.H., Wang, I.H.C., Khoo, L.P.: A product configuration analysis method for emotional design using a personal construct theory. Int. J. Ind. Ergon. **44**(1), 120–130 (2014)
10. Khalid, H.M.: Embracing diversity in user needs for affective design. Appl. Ergon. **37**(4), 409–418 (2006). Special Issue: Meeting Diversity in Ergonomics
11. McLachlan, G., Peel, D.: Finite Mixture Models. Willey Series in Probability and Statistics (2000)
12. Nagamachi, M.: Kansei engineering: a new ergonomic consumer-oriented technology for product development. Int. J. Ind. Ergon. **15**(1), 3–11 (1995)
13. Osgood, C.E., Suci, G.J., Tannenbaum, P.H.: The Measurement of Meaning. University of Illinois Press (1964)
14. Petiot, J.F., Yannou, B.: Measuring consumer perceptions for a better comprehension, specification and assessment of product semantics. Int. J. Ind. Ergon. **33**(6), 507–525 (2004)
15. Rossi, P.E., Allenby, G.M., McCulloch, R.: Bayesian Statistics and Marketing. Wiley, Hoboken (2005)
16. Schütte, S.: Engineering emotional values in product design: Kansei engineering in development. Ph.D. thesis, Institutionen för konstruktions-och produktionsteknik (2005)
17. Tibshirani, R.: Regression shrinkage and selection via the lasso. J. R. Stat. Soc. **58**(1), 267–288 (1996)
18. Wang, W., Li, Z., Tian, Z., Wang, J., Cheng, M.: Extracting and summarizing affective features and responses from online product descriptions and reviews: a Kansei text mining approach. Eng. Appl. Artif. Intell. **73**, 149–162 (2018)

19. Zhai, L.Y., Khoo, L.P., Zhong, Z.W.: A rough set based decision support approach to improving consumer affective satisfaction in product design. Int. J. Ind. Ergon. **39**(2), 295–302 (2009)
20. Zikopoulos, P., Eaton, C., et al.: Understanding Big Data: Analytics for Enterprise Class Hadoop and Streaming Data. McGraw-Hill Osborne Media, New York (2011)

The Impact of Online Reviews on Product Sales: What's Role of Supplemental Reviews

Hao Liu, Jiangning Wu[⊠], Xian Yang, and Xianneng Li

Dalian University of Technology, Dalian 116024, Liaoning, China
{liuhao0330, yangxian}@mail.dlut.edu.cn,
{jnwu, xianneng}@dlut.edu.cn

Abstract. As the new form of online reviews, supplemental reviews have attracted the attention of many scholars. Considering current studies do not take full consideration on the content of reviews, the study goes better maturely and thoroughly on all information in supplemental reviews. In detail, sentiments of supplemental reviews in terms of different features corresponding to product, service and logistics are quantitatively analyzed. Except for the sentiments of reviews, other important factors including the volume of supplemental reviews, price, and ratings are introduced into the designed log-linear regression model for estimation. To explore the impact of supplemental reviews on product sales, an empirical study is conducted. The selected product with high involvement is mobile phone covering 32 products. The related sales as well as initial reviews and supplemental reviews are crawled from tmall.com for experiments. The period of data is from July 5, 2018 to July 24, 2018. By regression analysis, the results show that the sentiments of product features in both initial reviews and supplemental reviews and the sentiments of logistics features in supplemental reviews have significant positive impact on product sales. The volume of supplemental reviews has a negative impact on product sales. Compared with initial reviews, the impact of sentiments of product features on sales in supplemental reviews is greater.

Keywords: Online reviews · Supplemental reviews · Sales
Sentiment analysis

1 Introduction

Nowadays, more and more people enjoy shopping online owing to its convenience. Different with the shopping in brick-and-mortar store, online shopping pays more attention to the word-of-mouth (WOM) of products, such as consumer opinions, user experience and user-generated reviews. It was reported that 82% of consumers would like to scan reviews before purchasing on the B2B/B2C platform, and 93% of consumers state that they could be influenced by online reviews to make buying decisions. In view of the importance of online reviews, many retailer websites provide a review platform for consumers to give ratings on products they have bought and write comments for evaluating products or services.

© Springer Nature Singapore Pte Ltd. 2018
J. Chen et al. (Eds.): KSS 2018, CCIS 949, pp. 158–170, 2018.
https://doi.org/10.1007/978-981-13-3149-7_12

To further collect more post-purchase opinions of consumers, retailer websites like tmall.com, ctrip.com, etc. permit consumers to appraise their purchased products or services again in a period of time. These new forms of reviews are named as supplemental reviews, which follow the initial reviews for the same product. Supplemental reviews are also focus on advantages or disadvantages about products, service of online retailers and logistics speed like initial reviews do, which provide more real consumption experience of consumers. In recent years, some scholars have studied the role of supplemental reviews. Shi et al. (2016) presented that there are many differences between initial reviews and supplemental reviews in terms of the volume, length, interval, and sentiment strength of reviews [1]. Wang et al. (2015) found that supplemental reviews play a more important role than initial reviews in aspect of perceived usefulness [2].

Moreover, some researchers have found that the supplemental reviews have an influence on product sales [3, 4]. Shi et al. (2018) proposed that positive sentiments of both supplemental reviews and initial reviews as well as the volume of initial reviews can positively affect product sales [3]. Lin et al. (2017) found that poor comments in supplemental reviews have stronger negative impact on product sales than those in initial reviews [4].

However, in these studies, they just consider the sentiments at the review level, whereas ignore the sentiments at the feature level. In general, the online review content covers feature evaluation in three aspects, such as product feature, service feature, and logistics feature.

Hence, in this paper, supplemental reviews are deeply studied with their volume and contents. And three issues are detailed addressed in the following sections including (1) Does the volume of supplemental reviews can affect product sales?, (2) Do the sentiments of features in reviews corresponding to product, service and logistics can influence product sales?, and (3) Are there different impacts of sentiments of features on product sales for both initial reviews and supplemental reviews?

The main contribution of our paper is to examine the different influences of sentiments of product features, service features and logistics features on product sales in initial reviews and supplemental reviews. In practical, the research results are very important for retailers to manage the word-of-mouth of products and further improve the sales.

2 Related Works

2.1 Online Reviews

Chatterjee (2001) firstly coined the concept "online review" [5]. After that, online reviews have been widely used and deeply studied in different views. Park and Kim (2008) claimed that online reviews are positive or negative descriptions of products given by purchased users or potential users [6]. They evaluate products or services in different ways, say subjective/objective way, praising/dispraising way, and direct/indirect way. Owing to the information asymmetry between sellers and buyers, online reviews have become one of the important factors for consumers to make buying decisions.

Much more research has found that online reviews have significant impact on product sales. For instance, Clemons et al. (2006) studied the impact of online reviews on beer sales [7]. Chevalier and Mayzlin (2006) have found the important effect of online reviews on sales [8]. Liu (2006) used box office data to study the impact of online reviews on sales [9]. Zhu and Zhang (2010) investigated the significant effect of online reviews on game videos sales [10]. These researches can help more and more retailers realize that online reviews as a kind of referring factor are too important to be ignored.

From different perspectives, ratings, volume of reviews and text content of reviews have been discussed by many scholars. Based on the analysis of online reviews from dangdang.com, Gong (2013) proved that the volume of reviews and ratings on products both have significant positive effects on sales, and the degree of effect will go down along the life cycle of the product [11]. Luan and Neslin (2009) focused on new product market, and their study presented that the volume of reviews has significant impact on the acceptance of newly released game [12]. As to the content of reviews, some studies found that opinionated reviews play a more important role than ratings do [9]. Actually, numerical attributes of reviews such as ratings and the volume of reviews are easy to be compared. Whereas review content (e.g. how is the quality of a product in details) is inherently subjective and difficult to be evaluated, it should be quantitatively analyzed by scoring its sentimental intensity.

2.2 Supplemental Reviews

With the rapid development of e-commerce, online reviews have been further enriched in the existing form, and the online review mechanism is constantly improved. Most retailer websites have provided a new way for consumers to submit their supplemental reviews. The so called supplemental reviews refer to the re-evaluation of the same product or service after the period of use for consumers. Supplemental reviews can objectively reflect the consumer's purchased experience. Potential consumers believe that supplemental reviews are more useful than initial reviews [1]. Comparing with initial reviews, supplemental reviews cover more information on products, for instance, the feature "durability" for a product is seldom contained in initial reviews. With the acquaintance for the purchased product, consumers would like to add more usage experience in supplemental reviews, which results in higher weight on supplemental reviews [2].

In recent years, supplemental reviews have gradually attracted the attention of scholars. Li et al. (2014) pointed out supplemental reviews can really reflect the post-purchase experience of consumer because he/she has more time to experience the product [13]. Comparing with initial reviews for products, supplemental reviews are rather different in terms of their volume, length and sentiment intensity [1]. Li and Ren (2017) performed an empirical study and results showed that the inconsistent content between initial reviews and supplemental reviews can affect consumers' willingness to spend [14].

However, the present studies only worked on the supplemental review itself in a qualitative way, and ignored the sentiment orientation of the supplemental review, which has been proved to have a significant impact on product sales. This motivates us to figure out this issue in a quantitative way and try to reveal the different impact degrees of two forms reviews on product sales.

2.3 Features in Online Reviews

In online reviews, the features of product, service and logistics are often mentioned by consumers. Taking mobile phone as an example, features on mobile phone itself can be the pixel of the lens, battery endurance, etc., features on service can be the attitude of customer service personnel, the reputation of the retailers, etc., and features on logistics can be the integrity of packaging, timeliness of express delivery, etc. The existing researches just concern the product features effect on product sales. For example, Archak et al. (2011) studied the impact of opinions towards product features on sales by means of the sentiment analysis tool [15]. Wang (2016) proposed a new extraction method based on information theory to distill opinions on product features, and by regression analysis for 20 features of the given product, they found that features opinion have significant impact on product sales [16].

In reality, most consumers would like to read reviews according to their own concerns. At this point, it is worth to study the impact of sentiments of all features in online reviews on product sales. However, previous studies only take the product feature into account whereas ignore the impact of other features on consumer purchasing decision. In this paper, the impact of sentiments of reviews at feature level on product sales is thoroughly studied.

3 Theoretical Model and Sentiment Analysis Method

3.1 Theoretical Model

The study explores the impact of supplemental reviews on product sales from two dimensions including the volume of supplemental reviews and their sentimental orientation. The volume of reviews reflects conformity effect, that is, the more people talk about the products, and the more people become informed about them, which in turn lead to higher sales. So there is no doubt the volume of reviews can influence product sales. As to the contents of reviews, they reflect certain sentiments, either positive or negative, or strong or weak. Sentiment orientation is consumers' perception of the product or service, which can intensively influence consumer's purchase decision.

To encourage consumers to share their post-purchase experiences, online retailer may post some reward policies. Studies suggest that the behavior of participating in initial reviews is strongly affected by the given reward policy [17]. Unlike initial reviews, posting supplemental reviews or not is not decided by reward policy, instead dominated by consumer's subjective will. From this point of view, supplemental reviews, especially the volume of supplemental reviews, represent the deeper awareness of consumer on the product or service provided by the online retailer comparing with initial reviews. In addition, supplemental reviews are the outcome after purchase and experience of the consumer, which can be the complement and correction of initial reviews. So the usefulness of supplemental reviews is higher than initial reviews, since supplemental reviews reflect the real situation of product or service. We can image that supplemental reviews will have a major impact on product sales.

Apart from the volume of online reviews, the contents, ad hoc the sentiments, of reviews should be concerned which have been proved to have a significant impact on product sales. The sentiments of reviews are often attached to the multiple features. In this study, product features, service features and logistics features are mainly concerned and meanwhile their different roles in sales are also investigated. By measuring the sentiments of these features, the impact of supplemental reviews on product sales can be quantitatively depicted. The theoretical model for this purpose is put forward, as showed in Fig. 1 below.

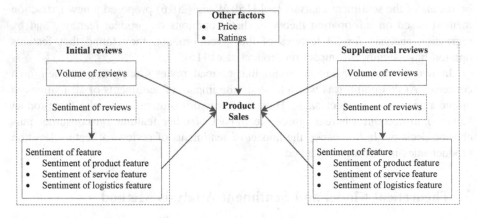

Fig. 1. Theoretical model

In the above model, two kinds of reviews are introduced, where the volume of reviews, sentiment of product feature, sentiment of service feature, and sentiment of logistics feature as well as ratings and product price are the independent variables, and product sales is the dependent variable. To realize the theoretical model, sales data of selected products and corresponding online reviews are collected at first. And then sentiment analysis is conducted on review contents. Finally, log-linear regression models are built for the empirical study.

For quantitative analysis by means of the proposed theoretical model, sentiment analysis should be done in advance.

3.2 Sentiment Analysis Method

For review sentiment analysis, three main tasks are involved, including feature identification, feature opinion identification and sentiment score calculation.

Feature Identification. Many techniques for identifying features mentioned in online reviews have been introduced in the last few years in text mining researches [15, 18, 19]. For our purpose, we can distill features from reviews by the following four steps.

Step 1. Data cleaning. Invalid data like reviews without useful information and repetitive reviews are identified and deleted.

Step 2. Word segmentation and tagging. Computer-based automatic word segmentation tool ICTCLAS (http://ictclas.nlpir.org/) is employed to segment the collected reviews and then tag the extracted terms. Next add the nouns and noun phrases into the candidate feature set and calculate the frequency of each term.

Step 3. Pruning. Delete the terms that are composed of only one Chinese character. Then use the thesaurus made by Harbin Institute of Technology to prune the candidate feature set, where synonymies have the same serial numbers. Next, replace the synonymies in the same set with a unique term and get the new candidate set with new term frequencies.

Step 4. Ranking. Rank candidate features terms by frequencies in descending order, and get the candidate feature list. Then recruit two judges who were asked to independently read the candidate feature list and delete the terms that are not features. After that, the final feature list can be obtained.

Feature Opinion Identification. To find out the opinion of feature which generally are adjective, adverbial and verb ones, word segmentation and part of speech tagging should be done at first. For this job, tool ICTCLAS is also employed to decompose review r into an array of words $\{w_1, w_2, ..., w_n\}$, where n is the total number of words. Then, a sentiment dictionary is utilized to match the word set, which was released by HowNet[1] on October 22, 2007. In the dictionary, there are seven word sets including positive (POS) words set, negative (NEG) words set, and five degree words sets corresponding to five levels of degree modifiers (MOD). Besides, both privative (PRI) word set and double privative (DB-PRI) words set are added into the sentiment dictionary. After matching, opinion words with different part-of-speech can be found out which are denoted by sw.

Sentiment Score Calculation. Once the opinion words have been tagged, the sentiment orientation can be analyzed quantitatively. For scoring the given opinion word, the following process is performed. At first, assigning the initial score of sw to 1 if it is a positive word based on the sentiment dictionary, otherwise assigning the initial score of sw to -1 if it is a negative word, and if neither of these is true, assigning the initial score to 0. Then, taking two words in front of the given sw, and multiplying the current score by -1 if they belong to PRI word set but not to DB-PRI word set. Finally, multiplying the weight of degree modifier if there are degree modifiers in front of the given sw. The pseudocode of sentiment scores calculation algorithm is shown in Fig. 2.

---

[1] http://www.keenage.com/html/e_index.html.

Algorithm 1 Review Sentiment Measurement

Input: Going into a loop to read a term w from word set

Output: Sentiment score s

1: $s \leftarrow 0$;

2: **if** $w \in$ POS **then**

3: $s \leftarrow 1$

4: **else if** $w \in$ NEG **then**

5: $s \leftarrow -1$;

6:　**end if**

7: **end if**

8: Initialize set of words with k words before w　$T = \{T_1, ..., T_k\}$;

9: **for** each T_i **do**

10:　**if** $T_i \notin$ DB-PRI && $T_i \in$ PRI **then**

11:　$s \leftarrow -1 * s$;

12:　**end if**

13: **end for**

14: Initialize set of words with m words before w　$T = \{T_1, ..., T_m\}$;

15: **for** each T_j **do**

16:　**if** $T_j \in$ MOD$_i$ **then**

17:　$s \leftarrow W_i * s$;

18:　**end if**

19: **end for**

20: **return** s;

Fig. 2. Pseudocode of sentiment score calculation algorithm.

Let Sr denote the sentiment score for feature r, we have

$$S_r = \frac{\sum\limits_{j=1}^{n} s_j}{\sum\limits_{j=1}^{n} |s_j|} \tag{1}$$

where S_r ranges from -1 to 1. And then we transfer the scale of S_r to the range of 0–1 by the min-max normalization, which way is also adopted in other studies [20]. $S_{rmin} = 0$ and $S_{rmax} = 1$, where a 0 score shows the least satisfaction, and a 1 score indicates the most satisfaction.

4　Empirical Study

4.1　Collected Data

The data in this paper come from China's largest B2C trading platform — Tmall. Unlike Amazon, JD.com and other B2C business platforms, which do not provide product sales information, Tmall's homepage shows the volume of reviews and monthly sales of products for all consumers at the same time. This provides a lot of reliable and effective data support for the study.

Some studies reveal that not all product purchase decisions involve an extensive information search because many products are of little importance to consumers. The concept of high-involvement and low-involvement products as a product classification is introduced to help distinguish consumer behaviors. Consumers conduct fewer information searches for lower-involvement products, but search extensively for high-involvement ones. The consumer often pays more attention to online reviews for the high involvement products than low involvement products. Thus, this paper chooses high involvement products to study. The mobile phone as a typical high-involvement product is then selected. The reasons for such selection are as follows. First, electronic products are typical high involvement products, and the sales volume of mobile phones is the highest in the electronics consumer market. Second, the mobile phone includes a variety of items, and market competition is fierce. Therefore, there are enough consumer reviews and sales datasets to support the study. For the selected mobile phones, we collect their IDs, price, volume of reviews, ratings, initial text reviews, supplemental text reviews and monthly sales for different 32 types with a 20-day period (from July 5, 2018 to July 24, 2018). After processing the collected data set, we extract three kinds of features and respectively identify their opinions from text reviews. The results are showed in Table 1.

Table 1. Statistical results for text reviews processing.

Feature type	No. of feature	No. of positive feature	No. of negative feature
Ini_product feature	9293	7658	1635
Ini_service feature	3351	2345	1006
Ini_logistics feature	4874	3828	1046
Sup_product feature	1467	962	505
Sup_service feature	458	228	230
Sup_logistics feature	224	163	61

According to Eq. (1), the sentiment scores of three kinds of features for both initial reviews and supplemental reviews can be obtained. The summary statistics of the collected data are shown in Table 2.

where $Month\_sales_{it}$ represents the monthly sales of product i at time t, Ini_volume_{it} represents the volume of initial reviews of product i at time t, $Price_{it}$ represents the price of product i at time t, $Rating_{it}$ represents the average rating of product i at time t, Sup_volume_{it} represents the volume of supplemental reviews of product i at time t, $Ini\_P\_sentiment_{it}$ represents the sentiment of product features of product i at time t in the initial reviews, $Ini\_S\_seniment_{it}$ represents the sentiment of the service features of product i at time t in the initial reviews, $Ini\_L\_sentiment_{it}$ represents the sentiment of logistics features of the product i at time t in initial reviews, $Sup\_P\_sentiment_{it}$ represents the sentiment of product features of the product i at time t in the supplemental

Table 2. Summary statistics of collected data.

Variables	Mean	Std. Dev	Max	Min
$Month\_sales_{it}$	894.734	1519.509	6316	27
Ini_Volume_{it}	898.970	778.388	2991	91
Sup_Volume_{it}	88.862	114.763	605	5
$Ini\_P\_Sentiment_{it}$	0.794	0.083	0.930	0.543
$Ini\_S\_Sentiment_{it}$	0.678	0.139	0.993	0.306
$Ini\_L\_Sentiment_{it}$	0.754	0.102	0.954	0.483
$Sup\_P\_Sentiment_{it}$	0.587	0.230	1	0.010
$Sup\_S\_Sentiment_{it}$	0.408	0.299	0.996	0
$Sup\_L\_Sentiment_{it}$	0.594	0.358	1	0
$Price_{it}$	2654.724	1945.740	9999	489
$Rating_{it}$	4.772	0.064	4.900	4.600

reviews, $Sup\_S\_seniment_{it}$ represents the sentiment of the service features of product i at time t in the supplemental reviews, $Sup\_L\_sentiment_{it}$ represents the sentiment of logistics features of the product i at time t in the supplemental reviews.

As we can see from Table 2, the number of supplemental reviews is less than the number of initial reviews. At the same time, sentiment scores of product features, service features and logistics features in supplemental reviews are lower than those in initial reviews. It suggests that consumers prefer to convey negative opinions in supplemental reviews.

The comparison of feature proportion in both initial and supplemental reviews is shown in the Fig. 3. The figure shows that the product features in supplemental reviews account for a higher proportion than initial reviews, which suggests that the supplemental reviews are more likely referring to the evaluation of product feature.

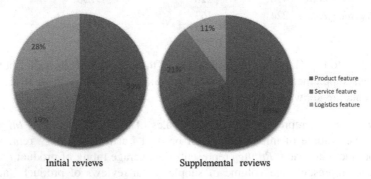

Fig. 3. Comparison of feature proportion in initial and supplemental reviews

4.2 Empirical Model

According to the proposed theoretical model, model I is designed to consider the impact of volume of supplemental reviews on product sales. The model employs the volume of initial reviews, product price and ratings as independent variables, which are given by Eq. (2):

$$ln(Month\_sales_{it}) = \alpha + \beta_1 * ln(Ini\_volume_{it}) + \beta_2 * ln(Price_{it})$$
$$+ \beta_3 * Rating_{it} + \beta_4 * ln(Sup\_volume_{it}) + \varepsilon \tag{2}$$

On the basis of model I, model II joins the sentiments of product features, service features and logistics features for both initial reviews and supplemental reviews. Model II is given by Eq. (3):

$$ln(Month\_sales_{it}) = \alpha + \beta_1 * ln(Ini\_volume_{it}) + \beta_2 * Ini\_P\_sentiment_{it}$$
$$+ \beta_3 * Ini\_S\_seniment_{it} + \beta_4 * Ini\_L\_sentiment_{it} + \beta_5 * ln(Sup\_volume_{it})$$
$$+ \beta_6 * Sup\_P\_sentiment_{it} + \beta_7 * Sup\_S\_seniment_{it}$$
$$+ \beta_8 * Sup\_L\_sentiment_{it} + \beta_9 * ln(Price_{it}) + \beta_{10} * Rating_{it} + \varepsilon \tag{3}$$

4.3 Results and Discussions

Using the log-linear regression functions of SPSS, the equations for the designed two models can be constructed. In order to exclude possible collinearity between variables, the regression method is chosen to be "Stepwise". The results for models I and II are shown in Tables 3 and 4, respectively.

Table 3. Estimation results for model I

Variables	Unstandardized coeff		Std. coeff	t	Sig.
	B	Std. error	Beta		
Constant	−18.274	2.673		−6.836	0.000
$Ln(Ini\_volume_{it})$	1.022	0.074	0.880	13.725	0.000
$Ln(Sup\_volume_{it})$	−0.608	0.059	−0.641	−10.306	0.000
$Ln(Price_{it})$	−0.650	0.051	−0.415	12.733	0.000
$Rating_{it}$	5.218	0.587	0.313	8.884	0.000
R^2	0.584				
Adjusted R^2	0.580				

From Table 3, we can see that R^2 of model I is 0.584. The estimated results show that the volume of initial reviews and ratings have a significant positive effect on product sales. But the price has a negative effect on product sales. This is consistent with existing research findings. However, the standardized coefficient of Sup_volume_{it} is -0.641, indicating that the volume of supplemental reviews has a negative impact on product sales. In other words, product sales will decrease with the increasing of the volume of supplemental reviews. This may be reasonable and can be explained as follows. As time goes by, consumers have a deeper understanding of the products they use. Poor user experiences will erupt in the form of negative reviews in supplemental reviews. As Table 2 showed, the sentiment score of each product feature in the supplemental reviews is lower than that of the initial reviews. Consumers are less likely to say negative things in their supplemental reviews than initial reviews. Meanwhile, traditional word-of-mouth studies show that negative reviews have a greater impact than positive ones [21, 22], and the more number of negative reviews, the less product sales. Hence, the volume of supplemental reviews with more negative sentiments will have a negative influence on sales.

Table 4. Estimation results of model II

Variables	Unstandardized coeff		Std. coeff	t	Sig.
	B	Std. error	Beta		
Constant	-14.672	2.739		-5.357	0.000
$Ln(Price_{it})$	-0.370	0.058	-0.253	-6.402	0.000
$Ln(Ini\_volume_{it})$	1.359	0.083	1.171	16.292	0.000
$Ln(Sup\_volume_{it})$	-0.892	0.063	-0.979	-14.078	0.000
$Sup\_P\_sentiment_{it}$	0.618	0.195	0.153	3.176	0.002
$Rating_{it}$	3.401	0.638	0.204	5.329	0.000
$Sup\_L\_sentiment_{it}$	0.480	0.121	0.164	3.973	0.000
$Ini\_P\_sentiment_{it}$	1.511	0.429	0.137	3.526	0.000
R^2	0.624				
Adjusted R^2	0.617				

Table 4 is the estimated results of model II, where R^2 of the model is 0.624. The standardized coefficients of $Ini\_P\_sentiment_{it}$ and $Sup\_P\_sentiment_{it}$ are 0.137 and 0.153 respectively. It suggests that compared with the initial reviews, the product features in the supplemental reviews have a greater positive impact on the product sales. Besides, the $Ini\_L\_sentiment_{it}$ is rejected by the model, and the sentiments of logistics features in the initial reviews have no significant influence on the product sales. But logistics features of the supplemental reviews have positive influence on product sales (say 0.164). Furthermore, both $Ini\_S\_sentiment_{it}$ and $Sup\_S\_sentiment_{it}$ are also rejected by the model, which suggests that evaluation of service feature has no significant effect on sales. These results indicate when consumers making purchasing decisions, they may care more about the evaluation of product feature and logistic feature than the service feature.

Therefore, retailers have to pay more attention to guiding and maintaining supplemental reviews from consumers. Some incentives can be taken to encourage consumers to make positive supplemental comments to enhance the word-of-mouth of product for retailers. What they can do is to encourage consumers to add supplemental reviews after initial reviews if they are satisfied with the product. Besides, retailers should pay more attention to the logistics and try to improve the service level to reduce the negative reviews from consumers when they add the supplemental reviews.

5 Concluding Remarks

As a complement to the initial reviews, the supplemental review is an improvement to the online reviews mechanism, which has an important impact on consumer purchasing decisions and product sales. In the study, we estimated the effect of supplemental reviews on product sales. The sentiments of supplemental reviews in terms of different features corresponding to product, service and logistics are quantitatively analyzed. The results of the study indicate that the volume of supplemental reviews with more negative evaluation has a negative impact on product sales. Besides, the sentiments of product features in both initial reviews and supplemental reviews, as well as logistics features in supplemental reviews also have a significant impact on product sales. Compared with initial reviews, the impact of sentiments of product features on sales is greater in supplemental reviews.

The current study still has certain limitations. Firstly, we only selected the mobile phone as the sample, and do not consider other products. In the future, we will compare and analyze the impact of consumer reviews on sales for different product categories such as high involvement products and low involvement products. In addition, the study only considered the volume of supplemental reviews and the sentiments of three kinds of features. In the future, the reviews can be deeply explored to analyze what content in product features and logistics features have a greater impact on consumers' purchase intentions.

Acknowledgements. We thank Tmall for providing plenty of products reviews. We also thank the project supported by Scientific and Technological Innovation Foundation of Dalian (2018J11CY009).

References

1. Shi, W., Gong, X., Zhang, Q., Wang, L.: A comparative study on the first-time online reviews and appended online reviews. J. Manag. Sci. **29**(04), 45–58 (2016). (in Chinese)
2. Wang, C., He, S., Wang, K.: Research on how additional review affects perceived review usefulness. J. Manag. Sci. **28**(3), 102–114 (2015). (in Chinese)
3. Shi, W., Wang, L., Sheng, N., Cai, J.: A comparative study into the impact of initial and follow-on online comments on sales. Manag. Rev. **30**(01), 144–153 (2018). (in Chinese)
4. Li, L., An, S.: The effect of additional difference on product sales. In: 19th China Management Science Academic Conference, vol. 25 (2017). (in Chinese)
5. Chatterjee, P.: Online reviews: do consumers use them? Adv. Consum. Res. **28**, 129–134 (2001)

6. Park, D.H., Kim, S.: The effects of consumer knowledge on message processing of electronic word-of mouth via online consumer reviews. Electron. Commer. Res. Appl. **7**(4), 399–410 (2008)
7. Clemons, E., Gao, G., Hitt, L.: When online reviews meet hyper differentiation: a study of the craft beer industry. J. Manag. Inf. Syst. **23**(2), 149–171 (2006)
8. Chevalier, J., Mayzlin, D.: The effect of word of mouth on sales: online book reviews. Nber Work. Pap.**43**(3), 345–354 (2006)
9. Liu, Y.: Word of mouth for movies: Its dynamics and impact on box office revenue. J. Mark. **70**(3), 74–89 (2006)
10. Zhu, F., Zhang, X.: Impact of online consumer reviews on sales: the moderating role of product and consumer characteristics. J. Mark. **74**(2), 133–148 (2010)
11. Gong, S., Liu, X., Zhao, P.: How do online consumer reviews impact sales? An empirical research based on online book reviews. China Soft Sci. **6**, 171–183 (2013). (in Chinese)
12. Luan, Y., Jackie, Y., Neslin, S.: The Development and Impact of Consumer Word of Mouth in New Product Diffusion. Social Science Electronic Publishing (2009)
13. Li, Z., Liu, R., Zhang, Y., Guan, H.: Statistical characteristics study on the Taobao's appended online review group based on complex networks. Soft Sci. **28**(08), 103–106 (2014). (in Chinese)
14. Li, Q., Ren, X.: Research on how conflictive additional reviews affect perceived helpfulness. J. Manag. Sci. **30**(04), 139–150 (2017). (in Chinese)
15. Archak, N., Ghose, A., Ipeirotis, G.P.: Deriving the pricing power of product features by mining consumer reviews. Manag. Sci. **57**(8), 1485–1509 (2011)
16. Wang, W., Wang, H.: The impact of feature opinions on purchase decision: sentiment analysis method of online reviews. Syst. Eng. Theory Pract. **36**(1), 63–76 (2016). (in Chinese)
17. Sun, Y., Dong, X., McIntyre, S.: Motivation of user-generated content: social connectedness moderates the effects of monetary rewards. Mark. Sci. **36**(3), 329–337 (2017)
18. Hu, M., Liu, B.: Mining opinion features in customer reviews. In: National Conference on Artificial Intelligence, vol. 69, pp. 755–760 (2004)
19. Ghani, R., Probst, K., Liu, Y., Krema, M., Fano, A.: Text mining for product attribute extraction. ACM SIGKDD Explor. Newsl. **8**(1), 41–48 (2006)
20. Yang, X., Yang, G., Wu, J.: Integrating rich and heterogeneous information to design a ranking system for multiple products. Decis. Support Syst. **84**, 117–133 (2016)
21. Pee, L.G.: Negative online consumer reviews: can the impact be mitigated? Int. J. Mark. Res. **58**(4), 545–568 (2016)
22. Song, X., Sun, X.: Review of consumer adoption of online reviews. J. Mod. Inf. **35**(01), 164–169 (2015). (in Chinese)

Social Media and the Diffusion
of an Information Technology Product

Yinxing Li and Nobuhiko Terui[(✉)]

Graduate School of Economics and Management, Tohoku University,
Sendai, Japan
dgod1028@gmail.com, terui@tohoku.ac.jp

Abstract. The expansion of the Internet has led to a huge amount of information posted by consumers online through social media platforms such as forums, blogs, and product reviews. This study proposes a diffusion model that accommodates pre-launch social media information and combines it with post-launch sales information in the Bass model to improve the accuracy of sales forecasts. The model is characterized as the extended Bass model, with time varying parameters whose evolutions are affected by the consumer's communications in social media.

Specifically, we construct variables from social media by using sentiment analysis and topic analysis. These variables are fed as key parameters in the diffusion model's evolution process for the purpose of plugging the gap between the time-invariant key parameter model and that of observed sales.

An empirical study of the first-generation iPhone during 2006 and 2007 shows that the model using additional variables extracted from sentiment and topic analysis on BBS performs best based on several criteria, including DIC (Deviance Information Criteria), marginal likelihood, and forecasting errors of holdout samples. We discuss the role of social media information in the diffusion process for this study.

Keywords: Bass model · Diffusion · Hierarchical Bayes model
Predictive density · Social media data · Text analysis · Sentiment analysis
Time varying parameter · Topic model

1 Introduction

The expansion of the Internet has led to massive information posted by consumers online through social media such as forums, blogs, and product reviews. This provides an opportunity for firms to know consumers' product expectations and evaluations without the need for a direct survey.

A growing number of studies have examined the influence of user-generated content in marketing. Lee and Bradlow (2011) have proved that customer reviews can complement existing methods for generating attributes used in marketing analysis by comparing expert guides and consumer surveys. Netzer et al. (2012) have utilized large-scale, consumer-generated data on the Web to understand consumers' top-of-mind associative network of products and the implied market structure insights. Moe

© Springer Nature Singapore Pte Ltd. 2018
J. Chen et al. (Eds.): KSS 2018, CCIS 949, pp. 171–185, 2018.
https://doi.org/10.1007/978-981-13-3149-7_13

and Trusov (2011) showed that when studying the effect of consumer's ratings, the potentially endogenous relation between sales and ratings must be considered. They also build a predict model and proof that ratings have strong influence to future sales. Tirunillai and Tellis (2011) used a naïve Bayes classifier and support vector machine to classify user-generated online content to positive news and negative news and incorporated this information into a financial econometric model to forecast stock returns. Ritesh et al. (2017) also build predictive model by using sentimental analysis by using news to predict stocks. Besides, Fan et al. (2017) incorporate social media information into diffusion model by sentiment analysis for the purpose of improving forecasts. Igarashi et al. (2018) also use Twitter comment data to forecasting sales of personal computer and shows that social media information, especially sentimental analysis and key words extracted by Lasso Regression have strong influence to future sales.

In this study, we use not only sales data but also user-generated online content (or online comments) to describe and forecast the diffusion process of a new product, where online WOM data is plugged into the model as covariates for affecting the change of key parameters over time. From the modeling perspective, our model is characterized as a diffusion model with a time-varying parameter. This parameter variation of the diffusion model has been discussed for several reasons; for example, as a competitive activity, changes in marketing practice, different segments adopting products at different times (Eliashberg and Chatterjee 986), specification and measurement errors (Putsis 1998), and aggregation and omitted variables (Sarris 1973; Judge et al. 1985). Our proposed models belong to the class of systematic variation models (Mahajan et al. 2000, Ch. 11) and they share the advantages with other time-varying parameter models in producing fewer forecasting errors, as was shown by Putsis (1998) and Xie et al. (1997). In addition, our models provide insights into the time variation of parameters to guide the transition. Our proposed model shares their purpose and encompasses their study in the sense of not only adding objective information separately extracted by topic analysis, but also providing insights through interpreting time variations of parameters for managerial implication.

We evaluate parameter estimates using a Bayesian approach, and our inference is exact in the sense of not relying on asymptotic theory. The predictive density is numerically evaluated to reflect the uncertainty of point forecasts in decisions, as recently discussed by Terui and Ban (2014) and Takada et al. (2015). This characteristic of inference is intrinsic to the new product diffusion process as it uses a limited number of data points.

In the next section, we briefly introduce the text analysis used in our study, i.e., sentiment analysis using a naïve Bayes classifier and topic analysis by LDA (Latent Dirichlet Allocation) model. In Sect. 3, we propose the models and explain the estimation procedure. The empirical application is reported in Sect. 4. We apply our model to the diffusion process of the first-generation iPhone by augmenting the information set with user-generated content from the BBS of this product.

2 Text Analysis of Social Media

2.1 Sentiment Analysis

We first generate the numeric information from text data by classifying users' comments into one of three comment categories: positive, negative, or neutral (no relation). In particular, the number of positive comments before launch reflects the expectations of potential customers that can lead to after-launch sales.

We use the naïve Bayes classifier for text analysis, which has been effectively applied to the marketing problem (Tirunillai and Tellis 2011). It is a simple probabilistic model based on the Bayes theorem, with independence assumptions between words, and it is well-recognized to show good performance in text analysis.

When the vector of words $\mathbf{x}$ in a comment is given, the posterior probability $p(C_k|\mathbf{x})$ of classifying it to C_k (category k, i.e., positive, negative, or neutral) is calculated by Bayes' theorem as $p(C_k|\mathbf{x}) \propto p(C_k)p(\mathbf{x}|C_k)$. $p(C_k)$ is the prior probability and can be defined by calculating the share of positive comments among all comments in the training data. $p(\mathbf{x}|C_k)$ is the likelihood, implying the probability that this comment with the vector of words $\mathbf{x}$ happens when it belongs to C_k under the assumption of independence of word, i.e., $p(\mathbf{x}|C_k) = \prod_{i=1}^{n} p(x_i|C_k)$. Then, we classify the comments using the value of $\hat{y}$ from the function below:

$$\hat{y} = \arg\max_k p(C_k) \prod_{i=1}^{n} p(x_i|C_k) \tag{1}$$

2.2 Topic Analysis

Next, we extract the "topics" from a collection of documents in social media using the latent Dirichlet allocation (LDA) model (Blei et al. 2003), which is well-established in natural language processing and applied in a variety of disciplines. The LDA model is based on the assumption that each document can be viewed as a mixture of various latent topics, where topics follow a multinomial distribution over words. Contrary to the fact that the naïve Bayes classifier assumes that one document only has one topic, LDA assumes that each document is a mixture of various topics.

More specifically, LDA is a generative model allowing sets of observations to be explained by unobserved groups, explaining why some parts of the data are similar. Denote $w_{d,i}$ as the i-th word in document d and $z_{d,i}$ the (latent) topic of the i-th word in document d. The model assumes that $w_{d,i}$ has a vocabulary (v) distribution in topic k that follows a multinomial distribution $\left(w_{d,i} \sim Multi(\phi_{z_{d,i}})\right)$ and $z_{d,i}$ follows topic distribution $\left(z_{d,i} \sim Multi(\theta_d)\right)$ in document d. Then, the model describes the probability that vocabulary v appears in document d and is represented as the sum of the products of topic distribution and vocabulary distribution over possible K ways:

$$p(v|d) = \sum_{k=1}^{K} p(v|k)p(k|d) = \sum_{k=1}^{K} \phi_{v,k}\theta_{k,d}. \tag{2}$$

In the LDA model, the most common method to estimate latent parameter $\mathbf{z}$ is to use Gibbs sampling. However, when there is a large volume of text data like in our study, Gibbs sampling requires a lot of time to sample the parameters. Then we employ a popular way known as "collapsed Gibbs sampling," which analytically uses the natural conjugate of prior distribution to integrate out $\theta_{k|d}$ and $\phi_{w|k}$.

3 Models

3.1 Diffusion Model with Social Media Information

We use the new product diffusion model by Bass (1969) as the base model and extend it in the way of incorporating social media information. Then, we assume that the potential market size (m) and imitator ratio (q) are changing over time and their dynamics are partially driven by temporal communications among potential users. On the other hands, we assume that the innovator ratio (p) is constant which means the innovators are not influenced by communications among other potential users.

We expect different roles for sentiment analysis and topic models. The sentiment analysis extracts emotional and rather subjective feelings of "like" (positive) or "dislike" (negative) from consumers' BBS communications. On the other hand, topic analysis involves objective factors based on consumers' expectations and evaluations before and after the launch of a new product and their responses to marketing activity.

We employ the empirical model of Srinivasan and Mason (1986), which uses a continuous form of expression for the difference of cumulative sales ($x_t - x_{t-1}$) to define the model as

$$y_t = m_t[F(t|p, q_t) - F(t-1|p, q_{t-1})] + \varepsilon_t \tag{3}$$

where the cumulative density is written by

$$
\begin{aligned}
F(t|p, q_t) &= \frac{1 - \exp\{-(p+q_t)t\}}{1 + \frac{q_t}{p_t}\exp\{-(p+q_t)t\}} \\
&= \frac{1 - \exp\left\{-\left(\frac{1}{1+\exp\{-p^*\}} + \frac{1}{1+\exp\{-q_t^*\}}\right)t\right\}}{1 + \left(\frac{1+\exp\{-p^*\}}{1+\exp\{-q_t^*\}}\right)\exp\left\{-\left(\frac{1}{1+\exp\{-p^*\}} + \frac{1}{1+\exp\{-q_t^*\}}\right)t\right\}}
\end{aligned}
\tag{4}
$$

and ε_t is assumed to follow a normal distribution $\varepsilon_t \sim N(0, \tau)$.

We assume that the dynamics of parameters are partly explained by extracted variables from social media on the grounds that they contain changes in consumers' emotions, expectations, and evaluations. We describe this mechanism using a hierarchical model for the parameters in addition to diffusion model (4). More specifically, for the appropriately transformed dynamic parameter vector $\theta_t = (m_t^*, q_t^*)'$, where $m_t^* = \log m_t$, $p^* = \log(p/(1-p))$, and $q_t^* = \log(q_t/(1-q_t))$, and covariate vector z_t (including constants and variables) by analyzing social media data. We define the hierarchical model as

$$\theta_t = G z_{t-1} + \eta_t \tag{5}$$

where z_{t-1} is a covariate vector constituted from social media data, η_t is the two-dimensional vector of error terms and assumed to follow a normal distribution $\eta_t \sim N_2(0, \Sigma)$, where $\Sigma = diag(\tau_1, \tau_2)$. That is, the models are canonically represented by hierarchical nonlinear regression models.

We denote the static Bass model as Model 1, where we set the covariate as $z_{t-1} = 1$. Then, the first model (Model 2) uses three quantities to describe the comments: total number of comments and numbers of positive and negative comments. We define the covariate as $z_{t-1} = (1, s_{t-1}, po_{t-1}, ne_{t-1})'$, where s_{t-1} means the number of comments in $t-1$. po_{t-1} and ne_{t-1} are, respectively, the numbers of positive and negative comments in $t-1$.

The second model (Model 3) is defined when the covariate comprises constant terms and extracted topics as $z_{t-1} = (1, T_{1t-1}, T_{2t-1}, T_{3t-1})'$, where T_{it-1} is the number of i-th topics at $t-1$. Although there are some approaches about how to select the number of topics, we assume three topics for simplicity. The third proposed model (Model 4) combines Models 2 and 3 by setting $z_{t-1} = (1, s_{t-1}, po_{t-1}, ne_{t-1}, T_{1t-1}, T_{2t-1}, T_{3t-1})'$.

The proposed models are characterized as the hierarchical regression model whose parameters evolve over time, synchronizing with temporal changes of variables constructed from social media communications at a previous time. Since the first column of coefficient matrix G is the vector of parameters of the static Bass model (Model 1), these models are nested and include the original Bass model as a special case when additional text information has no information on parameter evolutions in Eq. (5).

3.2 Posterior Density for Model Parameters

In terms of Eqs. (4) and (5), the model is canonically described as a hierarchical nonlinear regression model with time-varying parameters. We use a Bayesian MCMC method to estimate parameters since the procedure of hierarchical regression models has been well-established and the necessary conditional posterior densities are available in closed form, except the time-varying parameter $\{\theta_t\}$. Then we can proceed with relatively efficient computational steps by combining Metropolis–Hasting sampling for three key parameters, with Gibbs sampling for the other parameters, which are very standard method in Bayesian estimation. Rather than other traditional point estimation like EM algorithm, we use these two Bayesian sampling methods due to observe posterior density to evaluate our models.

In fact, the joint posterior density of model parameters is formulated by

$$
\begin{aligned}
p(\{\theta_t\}, p, \tau, G, \Sigma | \{y_t, t\}, \{z_t\}) &\propto p(\{\theta_t\} | \{y_t, t\}, p, \tau) p(\tau | \{y_t, t\}, \{\theta_t\}, p) \\
&\times p(p | \{y_t, t\}, \{\theta_t\}, \tau) \\
&\times p(G | \{\theta_t\}, \{z_t\}, \Sigma) p(\Sigma | \{\theta_t\}, \{z_t\}, G) \quad (6)
\end{aligned}
$$

where the right-hand side of first line of (6) means the product of conditional posterior density for parameters in the diffusion model (4) and the second line means those for hierarchical model (5).

The sampling scheme of MCMC for this model is as follows. Starting from the initial parameter values, once $\{\theta_t\}$ is generated, the posterior density of hierarchical models $p(G | \{\theta_t\}, \{z_t\}, \Sigma)$ and $p(\Sigma | \{\theta_t\}, \{z_t\}, G)$ are available in closed forms, i.e., normal and inverted gamma distributions with given hyper parameters. On the other hand, the likelihood function $p(\{y_t, t\} | \{\theta_t\}, p, \tau)$ in Eq. (4) is combined with prior density $p(\{\theta_t\} | G, \{z_t\}, \Sigma)$ from hierarchical model (5) to evaluate the conditional posterior density as

$$
p(\{\theta_t\} | \{y_t, t\}, \tau, G, \{z_t\}, \Sigma) \propto p(\{y_t, t\} | \{\theta_t\}, p, \tau) p(\{\theta_t\} | G, \{z_t\}, \Sigma) \quad (7)
$$

We employ Metropolis–Hasting sampling for this posterior density. When $\{\theta_t\}$ and p is given, the conditional posterior density $p(\tau | \{y_t, t\}, \{\theta_t\}, p)$ of the right-hand side of (6) is known as an inverted gamma distribution.

Finally, the posterior density of key parameters of the Bass model at the original scale is obtained by inverse transformation of $\theta_t = (m_t^*, q_t^*)$ to (m_t, q_t), i.e., $m_t = \exp(m_t^*)$, $q_t = 1/(1 + \exp(q_t^*))$ and $p = 1/(1 + \exp(p^*))$; then, we can evaluate the joint posterior density as

$$
p(\{m_t, p, q_t\}, G | \{y_t, t\}, \{z_t\}). \quad (8)
$$

4   Empirical Application

4.1   Data

We use the numbers for quarterly global sales of first-generation iPhones from June 2007 to September, 2008, as we considering iPhone as one of the most representative innovative technology products. These data were obtained from www.statista.com. Though it is an old case, 1st generation of iPhone is still a good case for our research considering the feature of Bass model.

As for social media information, corresponding to global sales data, we use "gs-marena" (http://www.gsmarena.com/), a well-known BBS for mobile phones, where users from all over the world put their comments regarding mobile phones in which they are interested. Users of this BBS can access information on topics for all phones and provide their own comments or discuss topics with other users. We extract social media text data on the first-generation iPhone and collect its sales data until the next-generation iPhone (iPhone 3G) is released. In the BBS of gsmarena, a new topic for a mobile phone is usually created when this phone is first announced to the public by the company. On January 9, 2007, Steve Jobs gave a presentation on the iPhone and a thread was created the following day. Each comment has three elements: user, date, and comment text. We extract date and comment text only because user information is not used in this study. A total of 8,121 comments uploaded between January 10, 2007 and November 24, 2007 are divided into two groups: 1,500 comments for training data and 500 comments for test data. The daily text data are converted to quarterly data and we use the first four quarters for estimating models while the last two quarters are kept for holdout samples.

4.2 Sentiment Analysis

A conventional sentiment analysis uses two categories—positive and negative comments—to classify comments. However, this BBS usually has many unrelated comments such as questions and discussions. Then we classify training data into three groups: positive, negative, and no relation. We confirmed that the no relation group improves the accuracy of classification.

We classify all 8,121 comments of training data and then test the accuracy using test data comprising 500 comments. The prior distribution and accuracy are given in Table 1. The prior distributions $p(C_k)$ for three categories are calculated by counting the number of positive comments in training data by interpreting each comment manually, i.e., by making a dictionary: 39% for positive, comments like "apple rules. The best innovation ever.", "this is genius.... I love my iPhone. it's the best thing that ever happened to me." are in this group. 33.1% for negative, comments like "very disappointing about the camera" and "8 gb is too small memory" are in this group. 27.8% for neutral, interrogative sentences like "does it have radio?", "anyone know what language the phone have, or only english?" are major in this group. The accuracy is defined as the ratio of the number of hits over the number of comments in the test data of 500 comments. We found 94.2% of positively predicted comments in test data to be truly positive, with hit rates of 90% and 84.7%, respectively, for negative and neutral predicted comments. This shows the high precision of our dictionary for sentiment analysis.

Table 1. Summary of Naïve Bayes classifier

Class	Prior (Training data)	Accuracy (Test data)
Positive	0.390	0.942
Negative	0.331	0.900
No relation	0.278	0.847

Figure 1 shows time-series plots for the numbers of positive and negative comments used in our study. The movements of these numbers are synchronized with those of sales with the lag of one period; thus, these can be leading indicators for sales.

Comments and Sales

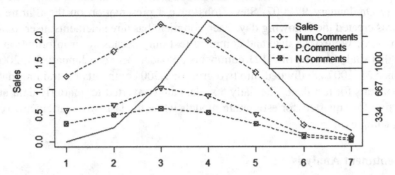

Fig. 1. Time-series plots of positive and negative comments

4.3 Topic Analysis

In the topic model, we set the number of topics based on following reasons. Firstly, considering only four time points are used as training data, more than 3 topics may have overfitting problem. Secondly, based on Blei et al. (2003), we compared perplexity among 1 to 3 topics, finally we define topic number as 3. The Bayesian Collapse Gibbs sampling algorithm is used to estimate the model. The number of $M = 4,000$ samples is used to evaluate posterior probability after discarding the previous 1,000 samples as a burn-in period. This computation needed a tremendously long time of about one week. Table 2 shows the top words for 3 topics, and the number next to each word refers to its rate of frequency in the document for each topic.

Table 2. Top words and their rate of frequencies for each topic

Topic1 (0.554)		Topic1 (0.224)		Topic1 (0.222)	
phone	0.0093	ur	0.0117	apple	0.0063
n95	0.0072	me	0.0081	mobile	0.0059
nokia	0.0071	can	0.0080	will	0.0058
good	0.0070	fone	0.0080	market	0.0058
but	0.0069	install	0.0080	contract	0.0054
better	0.0069	tell	0.0061	network	0.0051
people	0.0054	help	0.0061	europe	0.0049
camera	0.0053	thanks	0.0060	sim	0.0048
think	0.0041	bluetooth	0.0049	us	0.0047
like	0.0041	plz	0.0048	released	0.0041
really	0.0039	installer	0.0041	uk	0.0032
because	0.0038	files	0.0040	june	0.0032

First, the topic number in the table means the estimate of probability that topic k is in all D documents, i.e., the posterior mean $E\left(\theta_{k,\cdot} = \sum_d \theta_{k,d}/D\right)$. Topic 1 has the largest probability of 0.554 and is dominant compared with other topics with almost half the probability. Next, the top twelve words for each topic are given in their order of frequency. According to these classifications, we can easily characterize each topic. Topic 1 contains "phone," "n95" (Nokia cellphone), "nokia," "good," etc., which are used regarding reviews. Topic 2 includes the words "ur," "me," "install," "tell," "help," "thanks," "plz," and other words used in the context of discussions. Topic 3 contains "apple," "will," "market," "Europe," "us," "released," "uk," and other words in the context of marketing. Thus, we call Topic 1 "Reviews," Topic 2 "Discussion," and Topic 3 "Marketing."

Figure 2 shows time-series plots for the number of words in each topic and global sales data. The figure shows that the number of topics, especially Reviews and Marketing (Topics 1 and 3) leads to sales with a one-period lag and suggests that they could be leading indicators for accurate sales forecasting. In contrast, Discussion (Topic 2) is synchronized with sales.

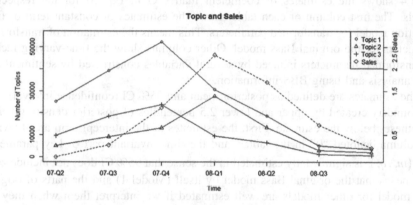

Fig. 2. Number of words in each topic

4.4   Model Comparison

The models were estimated by generated sample of $\theta_t^{(k)}, k = 1, \ldots, M$, and we used $M = 5,000$ samples for constructing the posterior density after discarding the previous 1,000 samples as the burn-in period. This required many iterations and almost 20 h for the MCMC sequence to converge for Model 4. Other models did not need such a high number of iterations. We confirmed their convergence using the Geweke's test (Geweke 1992), with a significance level of 95%. In the above, the non-informative diffuse prior was set for parameters.

Five models defined in previous section are compared based on three measures: log of marginal likelihood (LMD), deviance information criteria (DIC), and root mean squared errors (RMSE) of forecasts for holdout samples. The results are provided in Table 3.

First, the models with time-varying parameters perform significantly better than the static model (Model 1) in terms of respective criterion. This means that BBS contains useful information to describe the new product diffusion process. Among the dynamic

Table 3. Model Comparison

	RMSE	Log (ML)	DIC
Model 1	0.377	1.616	2.521
Model 2	0.105	2.997	2.509
Model 3	0.134	3.101	2.507
Model 4	0.027	4.589	2.235

models, the model with sentiment analysis (Model 2) is supported slightly better than the model with topics (Model 3). However, their combined model (Model 4) performs best.

4.5 Parameter Estimates

Table 4 shows the estimates of coefficient matrix G in Eq. (5) for the respective models. The first column of each table shows the estimates of constant term of time evolution model for transformed parameters. This means the estimation of transformed parameters for the original Bass model. Other columns show the time-varying factors of transformed parameters induced by several variables constructed by sentiment and topic analysis and using BBS information.

The estimates are defined as posterior mean and 95% CI (confidence interval) with the boundary created by upper and lower 2.5 percentiles of posterior density given in parentheses below the estimate. First, the estimates of the intercept term are shown in the column denoted "0" in the tables, and the time-invariant part of key parameters $\theta_t = \left(m_t^*, q_t^*\right)'$ is significantly estimated in the sense that 95% CI does not include zero. This means that the original Bass model by itself (Model 1) and the parts of original Bass model for other models are well estimated if we interpret them when they are inversely transformed. They drive a smooth orbit of sales and the mechanism of the original Bass model works for all models as an intrinsic part of the diffusion process. Second, three topics extracted by topic model affect all parameters not only when used solely, i.e., in the case of Model 3, but also together with sentiment variables, i.e., Model 4. Third, the numbers of total, positive and negative, comments from the sentiment analysis almost explain the changes of m_t^*, but they do not hold for q_t^* from the results of Models 2 and 4, and we could explain that imitators rely not on emotional but subjective factors such as review, discussion, and marketing by other people.

Next, we consider the result of the most supported model (Model 4) in more detail. The time transition equation of m_t^* has significant positive coefficients on all covariates and they induce a positive increase of m_t^* when they are increased. According to the magnitude of estimates due to the fact that the measurement scale is common in each category, the order of effectiveness is as follows: (i) Topic 2 (discussion) > Topic 3 (marketing) > Topic 1 (review) for topic models and (ii) positive comments > negative comments > number of comments.

Table 4. Parameter estimates

Model 1

	θ	num.comment	pos.comment	neg.comment	Topic1	Topic2	Topic3
m	2.065 *	-	-	-	-	-	-
	[2.026,2.044]	-	-	-	-	-	-
p	-3.015 *	-	-	-	-	-	-
	[-3.051,-2.924]	-	-	-	-	-	-
q	24.474 *	-	-	-	-	-	-
	[24.535,24.413]	-	-	-	-	-	-

Model 2

	θ	num.comment	pos.comment	neg.comment	Topic1	Topic2	Topic3
m	0.098	1.169 *	0.324	0.912 *	-	-	-
	[-1.156,0.345]	[0.716,1.335]	[-0.137,0.651]	[0.214,1.611]	-	-	-
p	-3.877 *	-	-	-	-	-	-
	[-4.726,-3.011]	-	-	-	-	-	-
q	0.527	0.446	0.677	1.137 *	-	-	-
	[-0.155,1.217]	[-0.110,0.893]	[-0.251,1.456]	[0.314,2.550]	-	-	-

Model 3

	θ	num.comment	pos.comment	neg.comment	Topic1	Topic2	Topic3
m	-0.008 *	-	-	-	0.303 *	0.878 *	-0.429 *
	[0.551,0.672]	-	-	-	[0.251,0.433]	[0.551,1.216]	[-0.522,-0.351]
p	-1.131 *	-	-	-	-	-	-
	[-1.450,-1.081]	-	-	-	-	-	-
q	-0.284 *	-	-	-	0.927 *	1.649 *	1.921 *
	[-0.368,-0.210]	-	-	-	[0.651,1.405]	[1.235,2.648]	[1.633,2.201]

Model 4

	θ	num.comment	pos.comment	neg.comment	Topic1	Topic2	Topic3
m	-0.453 *	0.069 *	0.795 *	0.198 *	0.315 *	0.845*	0.432 *
	[-0.551,-0.400]	[0.061,0.135]	[0.135,1.271]	[0.151,0.232]	[0.274,0.361]	[0.779,0.881]	[0.311,0.516]
p	-1.836 *	-	-	-	-	-	-
	[-1.870,-1.801]	-	-	-	-	-	-
q	-0.266 *	0.019	-0.027	-0.012	0.241	0.497*	0.957 *
	[-0.313,-0.201]	[-0.041,0.031]	[-0.051,0.047]	[-0.066,0.049]	[0.214,0.278]	[0.395,0.574]	[0.894,1.131]

Finally, the q_t^* equation has positive significant estimates of coefficient on three topic variables; however, there is no effective variable in sentiment analysis. This means that the change of imitator would be induced not by the subjective emotional factors in sentiment analysis, but rather by objective product evaluation through review and discussion in topic analysis.

4.6   Temporal Change of Key Parameters

The posterior density of key parameter estimates for Model 4, where the estimates of $\theta_t = \left(m_t^*, q_t^*\right)'$ and p^*, here $t = 1..., 4$ (estimates), and 5, 6 (forecasting) are inversely transformed to their original scales $\left(m_t, p_t, q_t\right)$ for the model interpretations. Most posterior densities are skewed by the form of log and logistic transformations. We share this skewness throughout the models and then define the estimates of the original key parameters by the median, which provides a more reasonable point estimate in the case of a skewed distribution.

4.7   Forecasting

Bayesian inference in this model constitutes unconditional predictive density. The predictive density for s-step ahead forecast y_{T+s} can be written by the model structure as

$$p(y_{T+s}|\text{Data}) = \int p(y_{T+s}|\theta_{T+s}, G, \Sigma)p(\theta_{T+s}|G, \text{Data})p(G|\Sigma, \text{Data})p(\Sigma|\text{Data})d\theta dG d\Sigma$$

$$(9)$$

where y_{T+s} is the s-step ahead forecast and θ_{T+s} is the corresponding time-varying parameter vector. The integration in Eq. (9) can be numerically evaluated by efficient Monte Carlo methods, i.e., by sequentially generating samples in addition to MCMC iterations for posterior density. That is, starting from some initial values of $\left(G^{(0)}, \Sigma^{(0)}\right)$, we take the steps: (i) $\Sigma^{(k)}$ is generated from $p(\Sigma|\text{Data})$; (ii) $G^{(k)}$ is generated from $p\left(G|\Sigma^{(k)}, \text{Data}\right)$, (iii) $\theta_{T+s}^{(k)}$ is generated from $p\left(\theta_{T+s}^{(k)}|G^{(k)}, \text{Data}\right)$ using Eq. (6); and (iv) $y_{T+s}^{(k)}$ is generated from $p\left(y_{T+s}^{(k)}|\theta_{T+s}^{(k)}, G^{(k)}, \Sigma^{(k)}\right)$ using Eq. (5). We note that when the diffusion model contains an explanatory variable of "time," the structural equation is easily updated by shifting T to $T + s$, without assuming scenarios for future explanatory variables, as is done by Takada et al. (2015).

Figure 3 shows the model fit for in-sample and out of sample data.

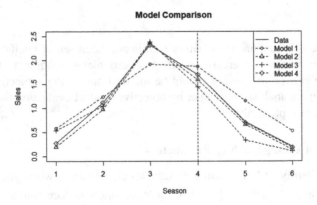

Fig. 3. In-sample and out of sample fit

Figure 4 shows the generated forecasts of respective models from the fifth and sixth periods, where in-sample fits from the first to fourth periods are also depicted and where the forecasts are defined as the mean of predictive density. The predictive densities for Model 4 are shown with observation by the x-mark in Fig. 4. They are well-defined and accommodate holdout observations in the center of density, implying

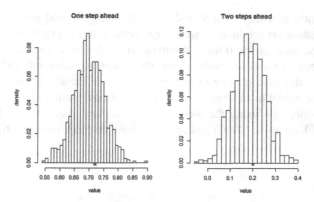

Fig. 4. Predictive density for model 4

that the forecast using predictive density has high precision. In addition, we can evaluate the predictive interval easily by evaluating percentiles of predictive density.

5 Concluding Remarks

The fusion of numeric structured data and unstructured text data is a challenging issue in big data analysis and it is also demanded in marketing research.

In this article, we proposed time-varying diffusion models to accommodate social media information. These models belong to the class of systematic variation models and provide useful insights on parameter variations, where we enlarge the information set regarding the diffusion process using product-related BBS text data from before and after the launch of a new product. We use this information based on the recognition that communications in BBS reflect changes in consumer expectations before launch as well as changes in product evaluations of not only the product itself but also the marketing activity and its competitive products. In particular, the communications among potential customers waiting to launch innovative IT products used in our study contain a sort of proxy variable for consumers' expectations before launch, changes in perception and evaluation after launch.

Our proposed models contain additional variables constituted from BBS text data by applying two approaches for analyzing text data, i.e., sentiment analysis and topic analysis. These variables are used as covariates to explain parameter temporal transitions. These analytical techniques are expected to extract subjective emotional variables and evaluation-based objective variables in BBS, respectively. The empirical study showed that these additional variables lead to an improvement in the model fit and precision of forecasting by filling a gap between smooth transitions of sales generated by a static diffusion model and realized sales, and they provide the roles of constructed variables in text analysis for the change in model parameters. For example, both of the emotional sentiment variables, rather than objective topic variables, have positive effects on market potential; on the other hand, topic variables affect the potential market and imitator with reasonable interpretation while sentiment variables do not affect the change of imitator transition. We also showed that the proposed model with the augmented information set produces a great improvement in the precision of forecasting.

IT products, such as the iPhone, continue to evolve and, together with growing social media networks, due to involve newer data including sales data and social media data and detect dynamic change of influence of social media, we can consider the extension of our model to successive product generations, including second- and third-generation products, by using the models of Norton and Bass (1987), Mahajan and Muller (1996), Kim et al. (2000) and others. Future research can investigate into this problem.

References

Bass, F.M.: A new product growth for model consumer durables. Manag. Sci. **15**(5), 215–227 (1969)

Blei, D.M., Ng, A.Y., Jordan, M.I.: Latent Dirichlet allocation. J. Mach. Learn. Res. **3**, 993–1022 (2003)

Eliashberg, J., Chatterjee, R.: Stochastic issues in innovation diffusion models with stochastic parameters. In: Mahajan, V., Wind, Y. (eds.) Innovation Diffusion Models of New Product Acceptance, pp. 151–203. Ballinger Publishing Co., Cambridge (1986)

Fan, Z.P., Che, Y.J., Chen, Z.Y.: Product sales forecasting using online reviews and historical sales data: a method combining the bass model and sentiment analysis. J. Bus. Res. **74**, 90–100 (2017)

Geweke, J.: Evaluating the accuracy of sampling-based approaches to the calculation of posterior moments. In: Bernardo, J.M., Berger, J.O., Dawid, A.P., Smith, A.F.M. (eds.) Bayesian Statistics 4, pp. 169–193. Oxford University Press, Oxford (1992)

Igarashi, M., Li, Y.X., Ishigaki, T., Terui, N.: Sales forecasting model using word-of-mouth on Twitter and empirical analysis. Distrib. Inf. **49**(6), 57–70 (2018). (in Japanese)

Judge, G.C., Griffiths, W.E., Hill, R.C., Lutkepohl, H., Lee, T.C.: The Theory and Practice of Econometrics, 2nd edn. Wiley, New York (1985)

Kim, N., Chang, D.R., Shoker, A.D.: Modeling inter-category dynamics for a growing information technology industry: the case of the woreless telecommunications industry. Manag. Sci. **46**(4), 496–512 (2000)

Lee, T.Y., BradLow, E.T.: Automated marketing research using online consumer reviews. J. Mark. Res. **48**(5), 881–893 (2011)

Mahajan, V., Muller, E.: Timing, diffusion, and substitution of successive generations of technological innovations: the IBM mainframe case. Technol. Forecast. Soc. Chang. **51**(2), 213–224 (1996)

Mahajan, V., Muller, E., Wind, Y. (eds.): New Product Diffusion Models. Kluwer Academic Press, Boston (2000)

Moe, W.W., Trusov, M.: The value of social dynamics in online products ratings forums. J. Mark. Res. **49**(3), 444–456 (2011)

Netzer, O., Feldman, R., Goldberg, J., Freskko, M.: Mine your own business: market-structure surveillance through text mining. Mark. Sci. **31**(3), 522–543 (2012)

Norton, J.A., Bass, F.M.: A diffusion theory model of adoption and substitution for successive generations of high-technology products. Manag. Sci. **33**(9), 1069–1086 (1987)

Putsis, W.P.: Parameter variation and new product diffusion. J. Forecast. **17**, 231–257 (1998)

Ritesh, B.R., Chethan, R., Harsh, S.J., Sheetal, V.A.: Stock market prediction using news articles. J. Emerg. Technol. Innov. Res. **4**(3), 153–155 (2017)

Sarris, A.H.: A Bayesian approach to estimation of time varying parameter regression coefficients. Ann. Econ. Soc. Meas. **2**(4), 501–523 (1973)

Srinivasan, V., Mason, C.H.: Nonlinear least squares estimation of new product diffusion models. Mark. Sci. **5**(2), 169–178 (1986)

Takada, H., Saito, K., Terui, N., Yamada, M.: The ubiquitous model for dynamic diffusion of information technology. Discussion Paper of DSSR, No. 35, Graduate School of Economics and Management, Tohoku University (2015)

Terui, N., Ban, M.: Multivariate structural time series models with hierarchical structure for over-dispersed discrete outcome. J. Forecast. **33**(5), 376–390 (2014)

Tirunillai, S., Tellis, G.: Does online chatter really matter? Dynamics of user-generated content and stock performance. Mark. Sci. **31**(2), 198–215 (2011)

Xie, J., Song, M., Sirbu, M., Wang, Q.: Kalman filter estimation of new product diffusion models. J. Mark. Res. **34**(3), 378–393 (1997)

The Effect of Cognitive Trust on Team Performance: A Deep Computational Experiment

Deqiang Hu, Yanzhong Dang, Xin Yue[✉], and Guangfei Yang

Institute of System Engineering, Dalian University of Technology,
Dalian 116024, China
xinyueyrx@mail.dlut.edu.cn

Abstract. This paper's purpose is to investigate the formation patterns of cognitive trust, the mechanisms by which it functions, and the characteristics of its influence on team performance. Toward this end, we present herein a deep computational experiment. We argue that a knowledge-intensive team is a complex adaptive system and that knowledge transfer in interpersonal interaction mediates between cognitive trust and team performance. Agent-based artificial teams, as a possible alternative form of real teams, are built in a computer, which acts as an experimental laboratory for investigating team activities. In particular, the modeling deeply penetrates internal psychological activities. A deep computational experiment is conducted under different internal and external conditions for the artificial team, yielding the following results. (1) Cognitive trust contributes to better team performance, while negative cognitive trust leads to worse team performance. (2) Simple and moderate tasks improve the formation of positive cognitive trust, while difficult tasks increase the formation of negative cognitive trust. The study method and findings presented herein are appropriate for other studies focusing on psychological effects on team, laying the foundations for new ideas for studying team building and team development.

Keywords: Cognitive trust · Team performance
Deep computational experiment · Knowledge transfer · Interpersonal interaction

1 Introduction

Faced with the increasing demands of complex competitive environments, organizations are looking to coordination and teamwork as means of resolving challenges, both large and small, across hierarchical levels [1]. In particular, the use of knowledge-intensive team (hereafter just "team") [2] has become an ideal model of work and working behavior for organizations [3]. Knowledge has emerged as the most strategic and significant resource of teams [4]. The transfer of knowledge [5], thus, forms not only the foundation of cooperation among team members but also an important measure to improve team performance [6]. Teams do not make full use of members' knowledge if those members fail to share and integrate the unique knowledge they each possess [7]. This requires organizations and researchers to focus more closely on

© Springer Nature Singapore Pte Ltd. 2018
J. Chen et al. (Eds.): KSS 2018, CCIS 949, pp. 186–200, 2018.
https://doi.org/10.1007/978-981-13-3149-7_14

processes such as interpersonal interaction that can support team members' effective knowledge transfer [8]. However, formal organizational rules are insufficient to ensure successful conduct of knowledge transfer. Most important for fostering knowledge transfer between team members [9], and the related synergy, is the existence of a climate of cognitive trust in interpersonal interaction [3].

Cognitive trust plays an important role in sharing interpersonal knowledge and is positively correlated with team performance [10]. In interpersonal interaction, a person who asks others for help may feel vulnerable. Cognitive trust can reduce this fear by creating an atmosphere that facilitates interpersonal interaction and by eliminating undesired and opportunistic behavior [9]. Cognitive trust may help to unlock members' potential by instilling greater self-confidence in their abilities to perform effectively, in addition to creating conditions in which members feel comfortable expressing differences in ways that enable the team to better learn from experiences and to identify more creative task strategies [11, 12]. Therefore, knowledge transfer in interpersonal interaction is only likely to occur in an atmosphere of cognitive trust in which people are of reliability and dependability [10].

Although cognitive trust is widely supported as an important antecedent of team performance [3, 13], little research has examined how cognitive trust between team members might actually be built along with teams' development and evolution [14]. Most researchers capture cognitive trust at static time points [15]. Missing from existing research is an examination of how cognitive trust evolves over time and the implications for team performance [16]. Simply stated, that static perspective does not stipulate the mechanisms by which individual-level action affects team-level outcomes [17].

Our primary research objectives are to explain how cognitive trust operates at the individual level of analysis, how it is related to knowledge transfer in interpersonal interaction, and, particularly, the mechanisms by which this inherently individual-level phenomenon translates into a team-level outcome: performance.

This paper presents a deep computational experiment to explore the effect of cognitive trust on team performance. The "deep" means that we develop herein a model to deeply probe internal psychological activities and focus on internal mechanism. The team is viewed as a complex adaptive system (CAS) [18], modeled as an artificial team using agent techniques in a "bottom-up" fashion. The two forms of the artificial team are regarded as possible alternatives of real teams. We utilize innovative computing technologies, employing a computer as an experimental laboratory for investigating team activities, such as knowledge transfer in interpersonal interaction, that are influenced by cognitive trust. We aim to reveal the mechanisms of how cognitive trust affects team performance.

2 Modeling

2.1 Team Model

As a complex adaptive system, the team's input is tasks and its output is team performance. Team members are adaptive agents whose behavior is influenced by cognitive trust and who adjust their behavior according to their environment. The team's target is to complete its tasks, which are the driving force behind the team's development [19]. Interpersonal interaction is team's main behavior through which members complete the tasks. The occurrence of knowledge transfer in interpersonal interaction contributes to team members' task completion [20]. Moreover, cognitive trust, as an extremely important psychological variable, develops through interpersonal interaction and, in turn, has a crucial effect on decision making in knowledge transfer. Thus, interpersonal interaction leads to complex interpersonal relationships and makes the team a complex adaptive system, as depicted in Fig. 1.

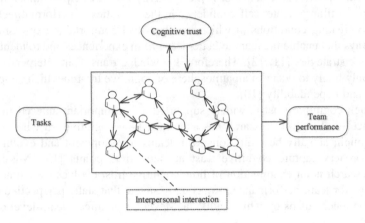

Fig. 1. Team model

Let $A = \{A_i | i = 1, 2, \ldots, m\}$ denote the set of team members, where A_i represents team member i, who is described as an agent. The number of team members is m, which satisfies $m \in \mathbf{N}^*$ and $m \geq 2$. We set cognitive trust that develops adaptively as the most important interaction relationship between team members.

Task Model. A task comprises several sub-tasks that are relatively independent. A task is denoted as T, defined as:

$$T = \{ST_i | i = 1, 2, \cdots, p\} \tag{1}$$

where ST_i represents the sub-task i. The number of sub-tasks is p, which satisfies $p \in \mathbf{N}^*$. For this study's purposes, experience, skill, and ability are collectively called knowledge, which is the only attribute of the task. A task's knowledge attribute determines the required types and level of knowledge. Thus, it is necessary to define a knowledge space to represent the knowledge.

The space $\mathbf{K}$ of q ($q \in \mathbf{N}^*$) dimensions denotes the knowledge space, where each dimension represents one type of knowledge. $\mathbf{K}$ is defined as:

$$K = \{K_1, K_2, \cdots, K_q\} \tag{2}$$

Sub-task ST_i's knowledge attribute is a vector in $\mathbf{K}$. It is denoted as $k(ST_i)$, defined as:

$$k(ST_i) = (k_1(ST_i), k_2(ST_i), \cdots, k_j(ST_i), \cdots, k_q(ST_i)) \tag{3}$$

Let $k_j(ST_i) \in [0, 1]$, $j = 1,2,...,q$ denote the amount of ST_i's K_j type of knowledge. Let $k_j(ST_i) = 0$ represent that ST_i does not involve the K_j type of knowledge. $k_j(ST_i) > 0$ represents that completing ST_i requires the K_j type of knowledge. $k_j(ST_i) = 1$ represents that completing ST_i requires the maximum amount of the K_j type of knowledge. There exists at least one j for which $k_j(ST_i) > 0$.

Team Performance. Since team development is driven by tasks, team performance can be evaluated by the degree to which tasks are performed within the allotted time. We define task-completion rate, task-completion efficiency, and comprehensive performance as the indexes for evaluating team performance. When a team has accepted T, the ST_i ($i = 1,2,...,p$) are allocated to A_j ($j = 1,2,..., m$); in general, $p \leq m$.

Task-completion rate is the degree to which a task has been completed. It is denoted as $CR(T)$, defined as:

$$CR(T) = \frac{N^{\text{finish}}(T)}{N^{\text{task}}(T)} \tag{4}$$

where $N^{\text{finish}}(T)$ is the number of tasks that the team has completed, and $N^{\text{task}}(T)$ is the number of tasks that the team has accepted, where $N^{\text{task}}(T) = p$ and $N^{\text{finish}}(T) \leq p$. Due to different knowledge levels and the inhomogeneity of knowledge distribution between members, the task may not be completed within the allotted time.

Task-completion efficiency is how fast a team can complete its tasks. It is denoted as $CE(T)$, defined as:

$$CE(T) = \frac{1}{N^{\text{task}}(T)} \sum_i \frac{t^{\text{allo}}(T) - t(ST_i)}{t^{\text{allo}}(T)} \tag{5}$$

Let $t^{\text{allo}}(T)$ denote allotted time interval. $t(ST_i) = t^{\text{inte}}(ST_i) + t^{\text{proc}}(ST_i)$, where $t(ST_i)$ denotes the time interval spent processing ST_i, $t^{\text{inte}}(ST_i)$ denotes the time interval spent interacting, and $t^{\text{proc}}(ST_i)$ denotes the time interval spent processing. Set $t^{\text{proc}}(ST_i) > 0$, $0 < t(ST_i) \leq t^{\text{allo}}(T)$. The smaller the $t(ST_i)$, the faster ST_i is processed and the higher the $CE(T)$.

Comprehensive performance is the weighted combination of task-completion rate and task-completion efficiency. It is denoted as $CP(T)$, defined as:

$$CP(T) = \omega \cdot CR(T) + (1 - \omega)CE(T) \tag{6}$$

where ω is the weight of $CR(T)$ and $1-\omega$ is the weight of $CE(T)$.

Team Cognitive Trust. Team cognitive trust denotes the average cognitive trust of all team members. It is denoted as $CT(\text{team})$, defined as:

$$CT(\text{team}) = \frac{1}{n(n-1)} \sum_{i=1}^{n} \sum_{j=1, j\neq i}^{n} CT(A_i, A_j) \tag{7}$$

where $CT(A_i, A_j)$ presents the cognitive trust of A_i for A_j. $CT(\text{team})$ varies with the team's development. Thus, at a given time point, $CT(\text{team})$ is the average value of $CT(A_i, A_j)$ at that time point, as discussed in greater detail in Sects. 2.2 and 2.4.

2.2 Individual Model

The individual model contains the attributes and behavior of an individual. Regarding the members of a real team, individual attributes include demographics, knowledge, cognitive trust, etc.; their behavior encompasses all types of actions involved in both work and life. For this study's purposes, individual attributes are simplified to comprise only knowledge and cognitive trust; individual behavior is also simplified to comprise only the actions required to process tasks.

Individual Attribute. *Knowledge Attribute.* $k(A_i)$ denotes A_i's knowledge and is a vector in knowledge space **K,** defined as:

$$k(A_i) = (k_1(A_i), k_2(A_i), \cdots, k_j(A_i), \cdots, k_q(A_i)) \tag{8}$$

Let $k_j(A_i) \in [0,1], j = 1,2,\ldots,q$ denote the amount of A_i's K_j type of knowledge. Let $k_j(A_i) = 0$ represent that A_i does not possess the K_j type of knowledge. $k_j(A_i) > 0$ represents that A_i has already mastered the K_j type of knowledge. $k_j(A_i) = 1$ represents that A_i possesses the maximum amount of the K_j type of knowledge.

Cognitive Trust Attribute. In this research, we use the definition of interpersonal trust proposed by Mayer et al. [21]: a party's willingness to be vulnerable to another party's actions based on the expectation that the other will perform an action important to the trustor, irrespective of the ability to monitor or control that other party. Expectation and reciprocation are important factors of cognitive trust [22]. Expectation determines how cognitive trust generates and reciprocation determines how cognitive trust develops.

We adopt the interpretation of cognitive trust (or distrust) as an interpersonal trust based on cognitive expectation of reliability and dependability [10]. Trust is based on the positive cognitive expectation that others will perform reliably and dependably, whereas distrust is based on the negative cognitive expectation that others will not act reliably and dependably or will even engage in potentially injurious behavior. In this study, cognitive trust contains both trust and distrust components, and is disparate in polarity and value.

An individual's cognitive trust attribute is a set that contains their cognitive trust in all the other members. A_i's cognitive trust attribute is denoted as $CT(A_i)$, defined as:

$$CT(A_i) = \{CT(A_i, A_j) | A_j \in A, j \neq i\} \tag{9}$$

where $CT(A_i, A_j)$ denotes the cognitive trust of A_i for A_j, $CT(A_i, A_j) \in [-1, 1]$. The range $[-1, 0]$ denotes distrust or negative cognitive trust; the range $[0, 1]$ denotes positive cognitive trust; and zero denotes that no cognitive trust exists. In general, $CT(A_i, A_j) \neq CT(A_j, A_i)$.

Individual Behavior. Individual behavior is the entire process from a member accepting a task until its completion or abandonment. This process comprises a set of ordered activities. Individual behavior is a coherent sequence of activities and judgement when an individual processes tasks. Individual task processing is shown in Fig. 2.

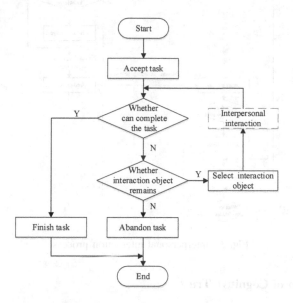

Fig. 2. Individual task processing

Interpersonal interaction is a more complex process, which will be discussed in detail in the following section.

2.3 Interpersonal Interaction Model

An interpersonal interaction involves two participants. The knowledge demander acts as the initiator and the knowledge supplier acts as the respondent. The communication between them is an interactive process. Both the initiator and respondent make multi-stage decisions that are influenced by cognitive trust. Either's present decisions are based on the previous decisions of the other. An interpersonal interaction includes two processes: negotiation and knowledge-transfer. Thus, interpersonal interaction is a process for not only knowledge transfer but also the formation of cognitive trust; it is also a process on which cognitive trust takes effect. Based on the study of Levin [6] and Alavi [1], the interpersonal interaction process is shown in Fig. 3.

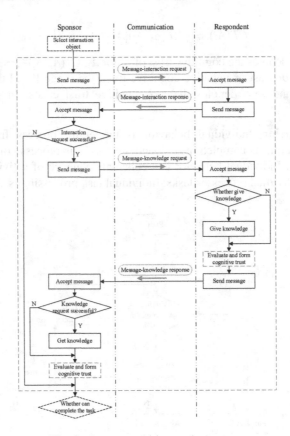

Fig. 3. Interpersonal interaction process

2.4   Formation of Cognitive Trust

Cognitive trust forms alongside the knowledge transfer process depicted in Fig. 3. The formation of cognitive trust contains three processes: (1) generation; (2) update; and (3) accumulation. We take the n-th interpersonal interaction between A_i and A_j, for example, to discuss the formation process of A_i's cognitive trust in A_j.

Generation of Cognitive Trust. In interpersonal interaction, a knowledge supplier's cognitive trust in the knowledge demander forms on the basis of cognitive expectation. The knowledge supplier infers from the knowledge demander that "you will treat me as I treat you" [10]. Offering knowledge to knowledge demander shows reliability and dependability of the knowledge supplier. So the knowledge supplier has the expectation that the knowledge demander will offer future help if and when needed by the knowledge supplier. Thereupon, the knowledge supplier's cognitive trust in the knowledge demander is generated. In other words, if the knowledge given by the supplier to the demander can satisfy the latter's need, then the supplier generally infers that a favorable attitude toward them is generated in the demander and believes that the demander will reciprocate. This favorable attitude generates positive cognitive expectation, which eventually converts to positive cognitive trust [14]. Otherwise, if the knowledge given by the supplier to the demander cannot satisfy the latter's need, then the knowledge supplier generally infers that an unfavorable attitude toward them is generated in the demander and believes that the demander will not reciprocate. This unfavorable attitude generates negative cognitive expectation, which eventually converts to negative cognitive trust [23].

The generation process of A_i's cognitive trust in A_j is shown in Fig. 4. A_i is the knowledge supplier and A_j the knowledge demander. When A_j needs knowledge, A_i generates an cognitive expectation of A_j by giving knowledge to them. This cognitive expectation is denoted as $CE_n(A_i, A_j)$, defined as:

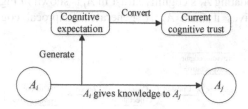

Fig. 4. Generation process of A_i's cognitive trust in A_j

$$CE_n(A_i, A_j) = \frac{K_n^{\text{give}}(A_i, A_j)}{K_n^{\text{need}}(A_j)} \tag{10}$$

where $K_n^{\text{give}}(A_i, A_j)$ denotes the knowledge amount that A_i gives to A_j, and $K_n^{\text{need}}(A_j)$ denotes the knowledge amount that A_j needs [$0 \leq K_n^{\text{give}}(A_i, A_j) \leq K_n^{\text{need}}(A_j), 0 < K_n^{\text{need}}(A_j) \leq 1, 0 \leq CE_n(A_i, A_j) \leq 1$]. When the amount of knowledge that A_i gives to A_j completely satisfies the latter's needs, A_i's cognitive expectation of A_j is maximized [$K_n^{\text{give}}(A_i, A_j) = K_n^{\text{need}}(A_j)$], which means that A_i strongly expects to receive knowledge from A_j in the next knowledge transfer. If A_i does not give any knowledge to A_j [$K_n^{\text{give}}(A_i, A_j) = 0$], A_i's cognitive expectation of A_j is minimized, which means that A_i does not expect to receive any knowledge from A_j in the next knowledge transfer.

This cognitive expectation will convert to current cognitive trust, which represents the cognitive trust given the most recent interaction and which eventually contributes to the accumulation of cognitive trust. Negative cognitive expectation converts to negative current cognitive trust, whereas positive cognitive expectation converts to positive current cognitive trust [23]. We make a proposition regarding the relationship between cognitive expectation and current cognitive trust. $CCT_n(A_i, A_j)$ denotes the current cognitive trust of A_i in A_j. If $0 \leq CE_n(A_i, A_j) < 0.5$, A_i has a negative expectation of A_j, expecting that their willingness to reciprocate is weak, $CCT_n(A_i, A_j)$ will be negative. If $CE_n(A_i, A_j) = 0.5$, A_i has a neutral expectation of A_j, A_i is unable to determine whether A_j will reciprocate; thus, $CCT_n(A_i, A_j)$ will be zero. If $0.5 < CE_n(A_i, A_j) \leq 1$, A_i has a positive expectation of A_j, expecting that their willingness to reciprocate is strong, $CCT_n(A_i, A_j)$ will be positive. $CCT_n(A_i, A_j)$ is defined as:

$$CCT_n(A_i, A_j) = \begin{cases} \log_{0.5}^{2-2CE_n(A_i,A_j)}, 0 \leq CE_n(A_i, A_j) < 0.5 \\ \log_2^{2CE_n(A_i,A_j)}, \ 0.5 \leq CE_n(A_i, A_j) \leq 1 \end{cases} \quad (11)$$

Update of Cognitive Trust. When an individual signals cognitive expectation toward another individual and that other reciprocates the cognitive expectation, cognitive trust spirals upwards. By contrast, when cognitive expectations are not reciprocated, cognitive trust spirals downwards and may even reach distrust [24] (Fig. 5).

The process of updating A_i's cognitive trust in A_j is shown in Fig. 7. When A_i needs knowledge and receives it from A_j, A_i generates reciprocal cognitive expectation,

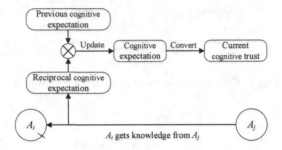

Fig. 5. Update process of A_i's cognitive trust in A_j

representing the extent to which A_i's cognitive expectation has been reciprocated by A_j. Cognitive expectation is updated by the mutual effect of reciprocal cognitive expectation and previous cognitive expectation, and eventually converts to current cognitive trust.

A_i's reciprocal cognitive expectation of A_j is denoted as $RCE_n(A_i, A_j)$, defined as:

$$RCE_n(A_i, A_j) = \frac{K_n^{\text{get}}(A_i, A_j)}{K_n^{\text{need}}(A_i)} \quad (12)$$

where $K_n^{\text{get}}(A_i, A_j)$ denotes the amount of knowledge that A_i gets from A_j, and $K_n^{\text{need}}(A_i)$ denotes the amount of knowledge that A_i needs [$0 \leq K_n^{\text{get}}(A_i, A_j) \leq K_n^{\text{need}}(A_i)$, $K_n^{\text{get}}(A_i, A_j) = K_n^{\text{give}}(A_j, A_i)$, $0 < K_n^{\text{need}}(A_i) \leq 1$, $0 \leq RCE_n(A_i, A_j) \leq 1$]. If $K_n^{\text{get}}(A_i, A_j) = K_n^{\text{need}}(A_i)$, then we set $RCE_n(A_i, A_j) = 1$, which indicates that the cognitive expectation is totally reciprocated. If $K_n^{\text{get}}(A_i, A_j) = 0$, then we set $RCE_n(A_i, A_j) = 0$, which indicates that the cognitive expectation is not reciprocated at all.

A_i updates their cognitive expectation as follows:

$$CE_n(A_i, A_j) = CE_{n-1}(A_i, A_j) + \frac{RCE_n(A_i, A_j) - CE_{n-1}(A_i, A_j)}{2} \qquad (13)$$

where $CE_n(A_i, A_j)$ denotes the n-th cognitive expectation of A_i for A_j, and $CE_{n-1}(A_i, A_j)$ denotes A_i's $(n-1)$-th cognitive expectation of A_j. Initially, $CCT_0(A_i, A_j) = 0$, so we set $CE_0(A_i, A_j) = 0.5$ based on function (11). If $RCE_n(A_i, A_j) < CE_{n-1}(A_i, A_j)$, then cognitive expectation decreases; if $RCE_n(A_i, A_j) = CE_{n-1}(A_i, A_j)$, then cognitive expectation remains constant; and if $RCE_n(A_i, A_j) > CE_{n-1}(A_i, A_j)$, then cognitive expectation increases. $CE_n(A_i, A_j)$ also converts to current cognitive trust, as per function (11).

Accumulation of Cognitive Trust. Cognitive trust accumulates through current cognitive trust, which develops through the variation of cognitive expectation. Thus, cognitive trust accumulates from previous cognitive trust and current cognitive trust. If the current cognitive trust exceeds previous cognitive trust, then cognitive trust increases; otherwise, cognitive trust decreases. When an individual has no cognitive trust in the other, they exert no tendentious attitude towards the other nor have any cognitive expectation of them. Thus, current cognitive trust has greater weight compared to previous cognitive trust in the accumulation of cognitive trust. When an individual has cognitive trust in another at the highest or lowest level, they exert a tendentious attitude towards the other and have high cognitive expectations. Thus, current cognitive trust carries less weight compared to previous cognitive trust in the accumulation of cognitive trust [25]. $CT_n(A_i, A_j)$ denotes the cognitive trust accumulated by A_i in A_j up to the n-th interpersonal interaction, and is defined as:

$$CT_n(A_i, A_j) = \begin{cases} CT_{n-1}(A_i, A_j) + \left[1 - CT_{n-1}(A_i, A_j)\right]CCT_n(A_i, A_j), & CT_{n-1}(A_i, A_j) > 0 \\ CT_{n-1}(A_i, A_j) + \left[1 + CT_{n-1}(A_i, A_j)\right]CCT_n(A_i, A_j), & CT_{n-1}(A_i, A_j) \leq 0 \end{cases}$$
$$(14)$$

where $CT_{n-1}(A_i, A_j)$ denotes the cognitive trust accumulated by A_i in A_j up to the $(n-1)$-th interpersonal interaction. $CT_0(A_i, A_j) = 0$.

2.5   Effect of Cognitive Trust

The team member who reciprocates not only hopes to gain material rewards but also expects to gain the cognitive trust of others whose behavior may not be predicted. The reciprocation can enhance cognitive trust, which is, in turn, the guarantee of reciprocation. Based on past interactions, cognitive trust makes individuals more tolerant of future uncertainty, more open, and braver [26]. In interpersonal interactions, cognitive trust influences individuals' decisions and behavior, especially in the knowledge transfer process.

Individuals are more willing to transfer knowledge to those in whom they have a high level of cognitive trust, given that those members are expected to reciprocate and better transfer knowledge in the future. Therefore, cognitive trust influences the amount of knowledge transferred between individuals. When A_i gives K_h type of knowledge to A_j, the knowledge amount $K_n^{\text{give}}(A_i, A_j)$ is defined as follows:

$$K_n^{give}(A_i, A_j) = \left[k_h(A_i) - k_h(A_j)\right] \frac{1 + CT_{n-1}(A_i, A_j)}{2} \tag{15}$$

where $k_h(A_i) > k_h(A_j)$. The more cognitive trust A_i places in A_j, the more knowledge A_i is willing to transfer to A_j, and vice versa. When A_j requests knowledge from A_i, the knowledge transfer occurs only when $k_h(A_i) > k_h(A_j)$. When $K_n^{give}(A_i, A_j) > K_n^{need}(A_j)$, let $K_n^{give}(A_i, A_j) = K_n^{need}(A_j)$.

Although cognitive trust is one of the most important factors in selecting the interaction object, space limitations preclude discussion of the selection process in relation to team task processing, in addition to its inclusion in the computational experiments.

3   Computational Experiments and Results

3.1   Purpose of Computational Experiments

The purpose of the computational experiments is to examine how cognitive trust affects team performance though different experimental designs and analyses. Cognitive trust forms in interpersonal interaction and, in turn, affects knowledge transfer therein, ultimately influencing team performance.

We design two types of artificial teams for comparative analysis. The first type is the "equivalent exchange" artificial team (hereafter "EE team"), in which agents do not possess the cognitive trust attribute and are rational decision makers. The second type is the "cognitive trust" artificial team (hereafter "CT team"), in which agents possess the cognitive trust attribute and are emotional decision makers. We also design two knowledge-transfer strategies: equivalent-exchange strategy and cognitive -trust strategy. In the former strategy, the knowledge transfer process is not affected by cognitive trust. The amount of knowledge that an agent gives must equal the amount that they receive. Only in this situation ($K_n^{need}(A_i) > 0$, $K_n^{need}(A_j) > 0$, and $K_n^{give}(A_i, A_j) = K_n^{give}(A_j, A_i)$) can knowledge transfer occur. In the latter strategy, the knowledge transfer process is affected by cognitive trust. The agent is not constrained by the equivalent exchange for each knowledge transfer and will instead reciprocate based on their long-term relationship with others. Thus, the amount of knowledge that an agent gives to others is based on cognitive trust ($K_n^{give}(A_i, A_j)$ is calculated as per function (15)). In the EE team, the equivalent-exchange knowledge-transfer strategy is adopted; in the AT team, the cognitive -trust knowledge-transfer strategy is adopted. By comparing the team performance of these two team types, we can examine the possible relationship among cognitive trust, interpersonal interaction, and team performance.

3.2   Design and Results of Computational Experiment

Design. We set up two groups: group A is the experimental group and group B is the control group. In group A, it comprises a CT team. In group B, it comprises an EE team.

Within a given team, the agents' respective knowledge attributes are heterogeneous. There exists at least an h, $h \in \{1,2,...,q\}$, where $k_h(A_i) \neq k_h(A_j)$. Across teams, the corresponding agents' knowledge attributes are identical. For any h, we have $k_h(A_i) \in$ [0.2, 0.8]. The task is divided into three norms: simple, moderate, and difficult. Each norm requires a unique knowledge level that respectively corresponds to [0.2, 0.4], [0.4, 0.6], and [0.6, 0.8]. Each norm consists of 30 tasks which arrive by batch. Let $m = 10$, $p = 10$, $q = 10$, $\omega = 0.5$, and $t^{allo}(T) = 40$. All experimental results are averages over 25 runs.

4   Results

Analysis. As shown in Fig. 6, CT(team) varies over task batch. CT(team) decreases as tasks become harder: CT(team) is positive when the team processes simple or moderate tasks, but negative when it processes difficult tasks.

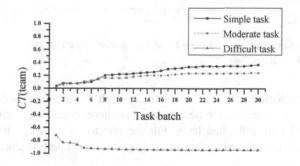

Fig. 6. Team cognitive trust in Group A over different batches across different task norms

As shown in Fig. 7, all three evaluation indexes of team performance decrease for groups A and B as the tasks become harder. When processing simple or moderate tasks, all three evaluation indexes for the CT team are relatively steady and at a high level. Conversely, all three evaluation indexes for the EE team are relatively unstable and are less than those for the CT team for most task batches. When processing difficult tasks, all three evaluation indexes fluctuate for groups A and B. However, for most batches, the overall trend is that all three evaluation indexes for group A are worse than those for group B. Based on our agent-based model, positive cognitive trust improve the knowledge transfer. So CT team has a better team performance under simple and moderate tasks for the continuous positive cognitive trust, while it has a worse team performance under difficult tasks for the continuous negative cognitive trust compared with EE team (as shown in Fig. 6)

As shown in Fig. 8, the results fall into two categories for all three evaluation indexes. The category next to the vertical axis is that in which CT(team) is negative and all three evaluation indexes are relatively low (results enclosed in full-line ellipse). The other category, which is far from the vertical axis, is that in which CT(team) is positive

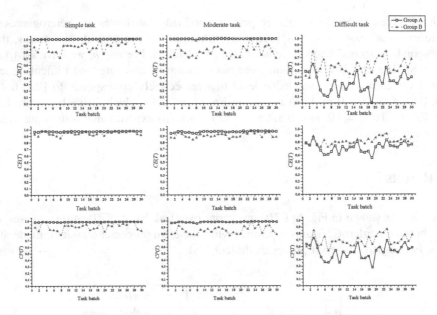

Fig. 7. Team performance over different batches across different task norms: comparison between CT and EE teams

and all three evaluation indexes are relatively high (results enclosed in dotted-line ellipse). Based on our agent-based model, positive cognitive trust improve the knowledge transfer in individual level. For the effect of knowledge transfer in inter-personal interaction, positive cognitive trust improves team performance in team level.

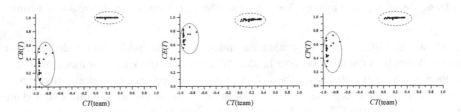

Fig. 8. Team performance across different levels of team cognitive trust

Why do the effects of cognitive trust vary with task norms despite agents' knowledge level matching that required by the tasks? When processing the simple or moderate tasks, an agent's knowledge generally exceeds what is required, so knowledge transfer can be easily conducted between agents. The knowledge transfer improves the formation of cognitive trust, in turn facilitating subsequent knowledge transfer and, ultimately, team performance. When processing the difficult tasks, some agents have insufficient knowledge to complete them and cannot get help from others. Here, the agent does not engage in reciprocal behavior. Cognitive trust, thus, decreases and sometimes even becomes distrust, which blocks the development of subsequent knowledge transfer and degrades team performance.

These results indicate that simple and moderate tasks improve the formation of positive cognitive trust, whereas difficult tasks increase the formation of negative cognitive trust. In addition, positive cognitive trust contributes to better team performance, while negative cognitive trust influences team performance deterioration.

5 Conclusion

In this paper, we have examined knowledge transfer in interpersonal interaction to explicate a mechanism for the connection between cognitive trust and team performance. A deep computational experiment is reported, which yielded the following findings. Simple and moderate tasks improve the formation of positive cognitive trust, whereas difficult tasks increase the formation of negative cognitive trust. In addition, positive cognitive trust contributes to better team performance, while negative cognitive induces team performance deterioration.

This work also has several managerial implications, which may enhance efforts to improve decision-making support for managers. Cognitive trust is a double-edged sword and should be duly handled by managers in team building. Sometimes, poor team performance may be caused by the lack of positive cognitive trust, rather than by insufficient knowledge, yet this factor is often ignored by managers.

This study still has the following limitations. First, the validity and efficiency of this paper's models need to be proven through further empirical studies. Second, the design of the computational experiments needs further consideration. Third, the study's comprehensiveness needs improvement.

To conclude, the deep computational experiment approach remains in its infancy, with extensive progress required before its systematization, scientification, and practicalization. We hope to enrich our research by further exploring the different effects of cognitive trust and affective trust on team performance, team building, and team development.

Acknowledgment. This work was partly supported by the National Natural Science Foundation of China under Grant No.71471028.

References

1. Alavi, M., Leidner, D.E.: Review: knowledge management and knowledge management systems: Conceptual foundations and research issues. MIS Q. **25**(1), 107–136 (2001)
2. Nordenflycht, A.V.: What is a professional service firm? Toward a theory and taxonomy of knowledge-intensive firms. Acad. Manag. Rev. **35**(1), 155–174 (2010)
3. Erdem, F., Ozen, J., Atsan, N.: The relationship between trust and team performance. Work Study. **52**(7), 337–340 (2003)
4. Yu, Y., Hao, J.X., Dong, X.Y., Khalifa, M.A.: Multilevel model for effects of social capital and knowledge sharing in knowledge-intensive work teams. Int. J. Inf. Manag. **33**(5), 780–790 (2013)

5. Argote, L., Ingram, P.: Knowledge transfer: a basis for competitive advantage in firms. Organ. Behav. Hum. Decis. Process. **82**(1), 150–169 (2000)
6. Levine, S.S., Prietula, M.J.: How knowledge transfer impacts performance: a multi-level model of benefits and liabilities. Organ. Sci. **23**(6), 1748–1766 (2012)
7. Nonaka, I.: A dynamic theory of organizational knowledge creation. Organ. Sci. **5**(1), 14–37 (1994)
8. Argote, L., Ingram, P., Levine, J.M., Moreland, R.L.: Knowledge transfer in organizations: learning from the experience of others. Organ. Behav. Hum. Decis. Process. **82**(1), 1–8 (2000)
9. Levin, D.Z., Cross, R.: The strength of weak ties you can trust: the mediating role of trust in effective knowledge transfer. Mana. Sci. **50**(11), 1477–1490 (2004)
10. McAllister, D.J.: Affect-based and cognition-based trust as foundations for interpersonal cooperation in organizations. Acad. Manag. J. **38**(1), 24–59 (1995)
11. Ng, K.Y., Chua, R.Y.J.: Do I contribute more when I trust more? Differential effects of cognition- and affect-based trust. Manag. Organ. Rev. **2**(1), 43–66 (2006)
12. Schaubroeck, J., Lam, S.S., Peng, A.C.: Cognition-based and affect-based trust as mediators of leader behavior influences on team performance. J. Appl. Psychol. **96**(4), 863–871 (2011)
13. Dirks, K.T.: The effects of interpersonal trust on work group performance. J. Appl. Psychol. **84**(3), 445 (1999)
14. Mayer, R.C., Davis, J.H.: The effect of performance appraisal system on trust for management: a field quasi-experiment. J. Appl. Psychol. **84**(1), 123–136 (1999)
15. Lewicki, R.J., Tomlinson, E.C., Gillespie, N.: Models of interpersonal trust development: Theoretical approaches, empirical evidence, and future directions. J. Manag. **32**(6), 991–1022 (2006)
16. Webber, S.S.: Development of cognitive and affective trust in teams. Small Group Res. **39**(6), 746–769 (2008)
17. Zaheer, A., McEvily, B., Perrone, V.: Does trust matter? Exploring the effects of interorganizational and interpersonal trust on performance. Organ. Sci. **9**(2), 141–159 (1998)
18. Holland, J.H.: Hidden order: How adaptation builds complexity. Addison Wesley, New York (1996)
19. Chae, S., Seo, Y., Lee, K.C.: Effects of task complexity on individual creativity through knowledge interaction: a comparison of temporary and permanent teams. Comput. Human Behav. **42**(SI), 138–148 (2015)
20. Majchrzak, A., More, P.H.B., Faraj, S.: Transcending knowledge differences in cross-functional teams. Organ. Sci. **23**(4), 951–970 (2012)
21. Mayer, R.C., Davis, J.H., Schoorman, F.D.: An integrative model of organizational trust. Acad. Manag. Rev. **20**(3), 709–734 (1995)
22. Mayer, R.C.: The reciprocal nature of trust: a longitudinal study of interacting teams. J. Organ. Behav. **26**, 625–648 (2005)
23. Lewicki, R.J., McAllister, D.J., Bies, R.J.: Trust and distrust: new relationships and realities. Acad. Manag. Rev. **23**(3), 438–458 (1998)
24. Elangovan, A.R., Auer-Rizzi, W., Szabo, E.: Why don't I trust you now? An attributional approach to erosion of trust. J. Manage. Psychol. **22**(1), 4–24 (2007)
25. Das, A., Islam, M.M.: Secured trust: a dynamic trust computation model for secured communication in multi-agent systems. IEEE Trans. Dependable Secur. **9**(2), 261–274 (2012)
26. Lewis, J.D., Weigert, A.: Trust as a social reality. Soc. Forces **63**(4), 967–985 (1985)

Link Prediction Based on Supernetwork Model and Attention Mechanism

Yuxue Chi[1,2] and Yijun Liu[1,2(✉)]

[1] Institutes of Science and Development, CAS, Beijing 100190, China
yijunliu@casipm.ac.cn
[2] University of Chinese Academy of Sciences, Beijing 100049, China

Abstract. To make full use of various types of data in link prediction, we proposed a link prediction method (SA, the abbreviation of supernetwork and attention mechanism) with two parts: information extraction and similarity measurement. Information is extracted on the basis of supernetwork for its multilayered, aggregative and other characteristics. In this part, we defined the operating unit for the flexibility and depth of information extraction. With the help of information extraction, we can get different types of subnetworks, which can be used in the similarity measurement. The similarity measurement part is inspired by the idea of attention mechanism: the allocation of attention might be different according to the difference of both subnetworks and nodes. After studying three types of relationships in the supernetwork, we proposed a similarity index (SimSA) combined three relationship types. To test the new method, we compared SA with famous CN and RA in the real data set of Douban, a popular social network site, and verified the application value of the new method.

Keywords: Link prediction · Supernetwork · Attention mechanism

1 Introduction

Network is an important tool for expressing various complex systems. As a result, complex networks are increasingly concerned by scholars of physics, biology, informatics and economics, and link prediction is an important part of the research of complex network. The goal of link prediction is predicting the existence of links between nodes, and it can be used to improve the efficiency of the experiment, infer the development of disciplines, study social relations and so on [1].

Data have accumulated exponentially during the era of information. Except data directly related to links, there are also various types of data about the attributes of nodes, we thought there exist latent features which can improve the efficiency of link prediction. Taking into account the diversity of data, how to organize such huge unstructured data and how to determine the contribution of different kinds of data are related to the optimization of link prediction. Super network, a promising tool for modeling, provides an idea to organize multidimensional data. It is regarded as a formalism for the modeling and analysis of complex decision-making in the

© Springer Nature Singapore Pte Ltd. 2018
J. Chen et al. (Eds.): KSS 2018, CCIS 949, pp. 201–214, 2018.
https://doi.org/10.1007/978-981-13-3149-7_15

information age [2]. Super-network possesses multileveled, multilayered, multidimensional, multi-attributed, congestive, and aggregative characteristics [3], which is beneficial to organize complex information. So the supernetwork is used as the framework, and studied how to extract valuable information based on the super network model.

For link prediction, the similarity is one of the most important concepts, for it is generally accepted that the link usually exists between two similar nodes. When using multiple types of data for similarity measurement, the specificity of each type needs to be considered. Attention mechanism can help to select the most pertinent piece of information [4]. Inspired by the idea of attention mechanism, for supernetwork, the attention to the node varies not only from node to node but also from subnetwork to subnetwork. In other words, their contribution to similarity measurement may be different. We provided a new link prediction method consisting of two parts, information extraction and similarity measurement, and the second part takes the differences and similarities of both intra-subnetwork and inter-subnetwork data into account. To verify the effectiveness of the method, we crawled the real data in Douban, a social platform which is famous for its rich hobby data, and compared our method with popular link prediction methods CN and RA.

Section 2 reviews related works of link prediction and super-network. Section 3 establishes the supernetwork model and introduces details of information extraction. Section 4 introduces the calculation of SimSA, while Sect. 5 applies it to an actual data set. The paper concludes with a discussion providing some suggestions for further research work.

2 Related Work

Link-prediction problem asks: to what extent can the evolution of a social network be modeled using features intrinsic to the network itself [5]. When it comes to applications, link-prediction is not only the prediction of existent yet unknown links but also future links, and the sources of the features can be summarized as two categories: node-based feature and topology feature. The topology feature is favored by its stability and accessibility. For example, based on common neighbor, one of the important topological information, Adamic et al. [6] made the famous AA indicator. Zhou et al. [7] proposed another useful indicator called RA, which is motivated by the resource allocation process taking place on networks. Proper non-topological features can be used to improve link prediction [8], especially cold-start link prediction [9].

Overall, link prediction algorithms can be divided into three categories. Similarity-based algorithm is regarded as the mainstreaming class of algorithms with good presentation as well as lower computational complexity. Lin [10] presented an information-theoretic definition of similarity, and in which the similarity is measured by the ratio between information needed to state the commonality and needed to fully describe the agents. Except AA index and RA index, similarity-based algorithms also include CN index [11], SimRank [12], significant path index [13], effective path index [14] and so on. The second is maximum likelihood based algorithm. For this kind of algorithm, a fundamental model containing network organization rules or parameters is usually required. And typical representative models includes hierarchical structure model [15] and stochastic block model [16]. The third can be regarded as methods based on a Markov chain or machine learning. And algorithms such as neural network have been applied [17]. This method can design specific algorithms according to different scenarios, so that it can achieve higher prediction precision, for example, the prediction of links in certain fields can be partly regarded as recommendation system [18]. But the obvious defects are the high computational complexity and non-universality.

There are also studies of link prediction based on the super network. The concept of supernetwork was firstly proposed by Sheffi [19] in 1985, and at that time supernetwork was used in urban transportation networks. In 2002, Nagurney [20] cleared the concept as networks above and beyond existing networks. Supernetwork can be defined by using a hypergraph [21] and regarding each network as a vertex [22]. No matter how supernetwork is defined, the two definitions are just subs of supernetwork [23]. After all, development and improvement are necessary to supernetwork, especially as a tool to help human study the real world. In fact, supernetwork has been widely used in travel problem [24, 25], supply chain [26, 27], knowledge network [28, 29] and so on. The diversity of supernetwork applications partly proves the practicality of supernetwork, but it also reflects a fact that people pay more attention to the applications of the network [30] and it is difficult to summarize a universal theory of supernetwork.

Explosive growth of data provides a favorable external environment of supernetwork. Liu et al. [31] proposed a super edge prediction algorithm with super triangle concept for the first time. In super link prediction, richer data means more useful information as well as more redundant information. Compared with link prediction in usual network, although link prediction in supernetwork can organize rich data, it's also confronted with more hindrance: how to recognize valuable information, how to weight the relationships between nodes of different types... In this paper, based on the advantages of supernetwork, we proposed a link prediction method that focuses on both information extraction and similarity measurement.

3 Information Extraction in Supernetwork Model

3.1 Supernetwork Model

The real world is organized by countless underlying rules, so the performance is both complex and varied. Meanwhile, the overloads of data give it a veil, which makes the exploration of underlying rules harder. To explore the real world, the network has been regarded as an important tool, but more and more networks of the real world cannot be described by the usual network tool. For example, the development of social network site (SNS) generates lots of different types of data about users, such as friends, fans, habits, locations... Obviously, it's difficult to organize all of these data with usual network tool, which cannot present the links between different types of data.

Compared with usual network tool, supernetwork with followed definition can help us solve the first question: how to organize these data efficiently?

Let $V = \{v_1, v_1, \ldots, v_n\}$ is a finite set, if

(1) $E_i \neq \Phi (i = 1, 2, \ldots, m)$;
(2) $\bigcup_{i=1}^{m} E_i = V$,

The binary relationship $H = (E, V)$ is a supernetwork. The vertexes are $v_1, v_1, \ldots, v_n$, and $E = \{e_1, e_2, \ldots, e_m\}$ is the edge set. $e_i = \{v_{i1}, v_{i2}, \ldots, v_{ij}\}$ $(i = 1, 2, \ldots, m; j = 1, 2, \ldots, n)$ are superedges.

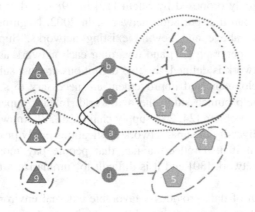

Fig. 1. Supernetwork for SNS, and the super-edge is linked by the same color lines.

Taking SNS as an example, each super-edge includes corresponding user's related information. Just like Fig. 1, nodes a, b, c, d represent users, and nodes 1, 2 ... 9 represent songs or books they like (pentagons represent songs). Super-edge $e_1 = \{1, 2, 3, a, 7\}$ means user a likes songs 1, 2, 3 and book 7. Super-edge $e_2 = \{1, 3, b, 6, 7\}$ means user b likes songs 1, 3 and book 6, 7. Supernetwork can efficiently sum these information up.

Nodes in supernetwork may contain different attributes, so we can divide them into different layers, which can be named as the **attribution layer**. With this idea, the example in Fig. 1 can be presented as Fig. 2.

In each attribution layer, edges could exist between nodes in the same layer, and we can name this kind of edges **intra-layer edges**. Limited by the data or research target, sometimes not all layers can contain intra-layer edges. Compared with link prediction in the usual network, the layer with edges that need to be predicted can be regarded as the **target layer**, and the edges can be regarded as the **target edges**. When we want to predict the relationship between users in SNS, the intra-layer edges in the user layer would be the target edges.

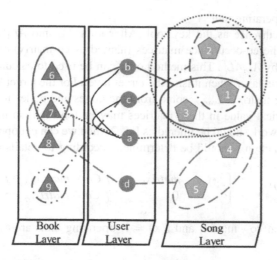

Fig. 2. Attribution layers of the example in Fig. 1

3.2   Information Extraction and Operating Unit

While in super-network, the main bridge between different layers is the super-edge, but super-edge is still far from our target, edges in the target layer. In other words, supernetwork model helps us organize data, and the next task is extracting useful information.

To extract information quickly and efficiently, we construct **operating unit** in each attribution layer, which can help us get information with different meaning. In the example, if two users like the same piece of music, they can be considered to have a connection. This process can be seen as the transformation from the super-edge to the target edge, so the key idea of operating unit is co-occurrence. And operating units in the example can be calculated by following two steps.

Step 1: Extract the adjacency matrix.
On the basic of supernetwork model, the adjacency matrix can be built as Eq. 1

$$AL_{i,j} = \begin{cases} 1 & if \ super - edge \ i \ passes \ point \ j \\ 0 & if \ super - edge \ i \ doesn't \ passes \ point \ j \end{cases} \tag{1}$$

The AL of the example as shown in Table 1.

Table 1. Matrix AL of the example.

	Book_1	Book_2	Book_3	Book_4	Book_5	Song_6	Song_7	Song_8	Song_9
SE_1	1	1	1	0	0	0	1	0	0
SE_2	1	0	1	0	0	1	1	0	0
SE_3	1	1	0	0	0	0	0	0	1
SE_4	0	0	0	1	1	0	1	1	0

Step 2: Get the operating unit

We choose co-occurrence as the key tool. After split AL into $AL_1, AL_2, \ldots, AL_n$ by attribution layer, the co-occurrence matrices (named as co-matrix) of different layers can be calculated by $AL_i AL_i^T$. This method only can be used to calculate the operating unit of an attribution layer when it isn't the target layer. For the target layer, AL_i should be replaced by the matrix AL_{target} come from the edges and nodes in the target layer.

But the numerical value in these matrices may be too high, so the co-occurrence matrices of subnetwork should be filtered as Eq. 2, then we can get operating units. The threshold t and function f should be determined according to details of examples.

$$OU_{i_{a,b}} = \begin{cases} f\left(Co - matrix_{i_{a,b}}\right) & if \ Co - matrix_{i_{a,b}} \geq t \\ 0 & if \ Co - matrix_{i_{a,b}} < t \end{cases} \tag{2}$$

When t = mean(co - matrix) and $f(x) = 1$, operating units are as Table 2.

Table 2. Operating units of the example.

OU_{user}	a	b	c	d		OU_{book}	a	b	c	d		OU_{song}	a	b	c	d
a	1	1	1	0		a	1	1	1	0		a	1	1	0	1
b	1	1	0	0		b	1	1	0	0		b	1	1	0	1
c	1	0	1	0		c	1	0	1	0		c	0	0	1	0
d	0	0	0	0		d	0	0	0	1		d	1	1	0	1

The operating unit represents relationships between nodes in the target layer via underlying rulers in attribution layers. The operating unit can be seen as the tool of information extraction, not only because it represents relationships from different aspects, but also because they can achieve deeper information extraction from richer aspects through simple operations, which can be summarized into two ways.

A. Within one attribution layer

The operating unit of target layer represents the connectivity between nodes through one neighbor. The information, such as the connectivity between nodes through the neighbor's neighbor, can be extracted by the matrix multiplication $AL_{target} OU_{target}$. To dig deeper for more information, we only need to repeat the multiplication.

For non-target attribution layer $AttL_i$, the similar multiplication is $OU_i OU_i$, which is mining information on the basic of relationship comes from common attributes between target nodes. The path to find the attribute grows with the operation length.

Whether in the target layer or non-target layer, the data can be further extracted by repeating the operation of matrix multiplication. The extraction can be named according to the number of operating units, for example, one operating unit corresponds to the 1-level extraction. As the length of operation increases, the value in new matrices may be too high, so the filtering operation mentioned earlier for high level extraction is also required.

B. Between different attribution layers

Considering that we need to predict edges in AL_{target}, we try to build the relationship between the target layer and non-target layers, so the extraction between different attribution layers mainly refers to the extraction between the target layer and non-target layers. For non-target attribution layer $AttL_i$, it can be achieved by the operation $AL_{target}OU_i$, which can be regarded as recommending nodes with the same attributes of node's neighbors in the target layer. Considering the order of multiplication can affect the result, it can be fixed during the operation. Another solution is updating elements in the matrix (E: $AL_{target}OU_i$) by the sum of corresponding elements about the main diagonal symmetry. Just like above, this operation can also be repeated.

With the help of operating units, we can extract information not only within one attribution layer but between different attribution layers. Considering that information extracted can be presented as network, they are named as subnetwork$_{Aik}$ and subnetwork$_{Bik}$ (Fig. 3), while A and B denote the first way and second way respectively, and i denotes the length of operation.

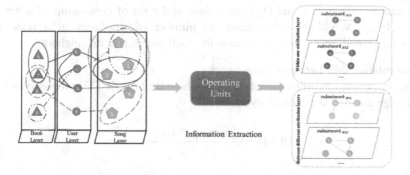

Fig. 3. Information extraction in the example

4 Link Prediction Method

After the extraction, the problem that comes with it is how to use different types of information for similarity measurement. Attention model can be considered as a resource allocation model, and we were inspired that the contribution of different types to similarity measurement should be different, which can be determined by the relationship between subnetworks and the target layer. For the information is extracted on the basic of supernetwork and the method is inspired by attention mechanism, the method can be shorted as SA. The first part of SA, information extraction, is introduced in Sect. 3, and the details of the next part are as follows.

According to the attention mechanism, people pay more attention to information that is valuable to people. According to this idea, it can be further expanded: valuable information may also be different for different people. This, combined with the super network, means that not only the contribution of different subnetworks to the similarity measurement is different, but the contribution distribution is likely to be different for different target nodes.

Through the above analysis, we know that the relationship is the key to this part, and there are mainly two objects: node and subnetwork. Besides the relationship between the subnetwork and the target layer, there are also two other relationships: the relationship between nodes, and the relationship between node and subnetwork.

A. The relationship between nodes

For the relationship between nodes, researchers have made a great contribution to the study of similarity in the usual network, and it's not difficult to use these indices of the usual network here.

B. The relationship between the node and subnetwork

For the relationship between node and subnetwork, a definition called dimension relevance [32] can be used (Eq. 3).

$$DR(v, D) = \frac{Neighbors(v, D)}{Neighbors(v, L)} \tag{3}$$

Equation 3 lets $v \in V$ and $D \subseteq L$ be a node and a set of dimensions of a network $G = (V, E, L)$. $Neighbors(v, D)$ means the number of neighbors of a node v in dimension D, and $Neighbors(v, L)$ means the total number of its neighbors.

C. The relationship between subnetworks

For the relationship between subnetworks, Michele Berlingerio [32] proposed an index called edge dimension connectivity as Eq. 4.

$$EDC(d) = \frac{|\{u, v, d\} \in E | u, v \in V|}{|E|} \tag{4}$$

EDC computes the ratio of edges of the network labeled with the dimension d. But this definition only considers the number of edges in different dimensions.

To achieve on our previous idea, we hope to propose a method to combine these three relationship types, or in other word, similarities, and then we can better distribute our attention by considering the differences between nodes and subnetworks. On the basis of the supernetwork, the similarity between target nodes a and b can be described as follows, while L is the number of subnetworks.

$$SimSA_{a,b} = \sum_{l=1,...,L}^{L} \frac{Similarity \ between \ a \ \& \ b}{Difference \ between \ a \ \& \ b \ in \ subnetwork \ l} \tag{5}$$

In Eq. 5, the denominator represents the difference between subnetworks, and the numerator represents the similarity between target nodes. The smaller the difference between two layers, the greater the contribution of the same numerator. This indicator extracts valuable information from multidimensional data. But until now, *SimSA* is just a concept. Because subnetworks mainly constructed by non-target layers, *SimSA* also needs information from the target network (OU_{tagert}) to prevent information offset. Combined with information extraction, the details of *SimSA* are as Eqs. 6 and 7.

$$simSA_{a,b} = C \times f_1 \left(\mathrm{OU}_{target_{a,b}} + d_1 \right) + SN_{a,b} \tag{6}$$

$$SN_{a,b} = \sum_{t=A,B} \sum_{i=1,...,I} \sum_{k=1,...,K} \frac{\log \left(subnetwork_{tik_{a,b}} + d_2 \right)}{\left| f_2 \left(Subnetwork_{i_{a,b}} - \mathrm{OU}_{target_{a,b}} \right) + d_3 \right|} \tag{7}$$

For *SimSA*, d_1, d_2, d_3 are usually greater than zero. f_1, f_2 can be determined according to the specific situation, while C is usually equal to the number of items in $SN_{a,b}$.

5 Computational Experiments

5.1 Data Set

As an open social platform, Douban contains rich data about hobby, so Douban is used as the experimental object in this paper. On the basic of python and MongoDB, we build a data acquisition system to collect user related data. In this experiment, in order to present the relationship between users as completely as possible, we collected around 200000 records about user' fans and friends to contrast the social network contains 236 users. We also crawled the information about hobbies, such as the books, movies and music, as the data indirectly related to the social network. Characteristic of the data set is in Table 3.

Table 3. Characteristic of the data set.

Category	Total category	Average category
Book	39458	202.3347
Movie	24067	383.6822
Music	6153	30.0508

The target layer in this data set can be displayed as Fig. 4, and prediction of edges in this layer are the target of link prediction.

5.2 Information Extraction

According to SA, we firstly build the super network. For there are three attributes (movie, book and music) used in the experiments, the supernetwork totally contains 4 attribution layers (another is the target layer). From Table 3, we know that book layer, movie layer and music layer contain 39458, 24067 and 6153 nodes respectively.

On the basic of super-network with attribution layers, we then calculate the operating units of three attribute layer, and $f(x) = 1$. The three operating units are as follows (Figs. 5, 6 and 7 can be drawn with VOSViewer, which is a visualization software developed by Van Eck and Waltman at the University of Leiden [33]).

Fig. 4. The social relationships between users in the data set

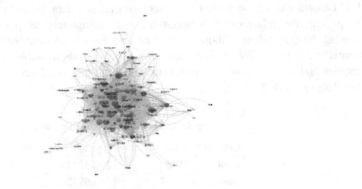

Fig. 5. The operating unit of book

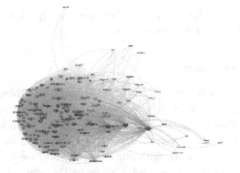

Fig. 6. The operating unit of movie

Fig. 7. The operating unit of music

From Figs. 5, 6 and 7, it's not difficult to find that operating units constructed with different attributions obviously have different structures. The contribution of the same user to the final result in different networks is different. In this case, Table 4 shows the extracted subnetworks. For ease of understanding, replace k in subnetwork$_{Aik}$ with words.

Table 4. Subnetworks of the data set.

Type	Subnetwork
Within one attribution layer	subnetwork$_{A1book}$, subnetwork$_{A1music}$, subnetwork$_{A1movie}$, subnetwork$_{A2book}$, subnetwork$_{A2music}$, subnetwork$_{A2movie}$
Between different attribution layers	subnetwork$_{B2book}$

5.3 Results

When calculating *SimSA*, $d_1, d_2, d_3 = 1$, $f_1(x) = x$, $f_2(x) = abs(x)$ and $C = 7$, for we extract 7 subnetworks.

In order to test the experimental results, we choose AUC and precision as evaluation indexes. Considering that CN and RA are both popular and useful indicator of link prediction, we compare our method with them.

Experimental results with different test set ratios are shown in Tables 5 and 6, and our method is represented by "SA".

Table 5. Table of AUC.

Proportion of test-set	10%	20%	30%	40%	50%	60%	70%	80%
CN	0.5085	0.5144	0.5196	0.5228	0.5245	0.5254	0.5224	0.5147
RA	0.5095	0.5152	0.5201	0.5232	0.5248	0.5255	0.5225	0.5147
SA	0.5011	0.5243	0.5461	0.5646	0.5832	0.6012	0.6161	0.6285

Table 6. Table of precision.

Proportion of test-set	10%	20%	30%	40%	50%	60%	70%	80%
CN	0.0111	0.0251	0.0282	0.0310	0.0292	0.0241	0.0269	0.0330
RA	0.0149	0.0406	0.0501	0.0552	0.0576	0.0596	0.0517	0.0330
SA	0.0174	0.0294	0.0394	0.0478	0.0495	0.0538	0.0549	0.0559

From the experimental results (Tables 5 and 6), we can find that

(1) For AUC, the performances of CN and RA are similar, and the performance of SA is better than both of them when the proportion of test-set is higher than 10%. For the precision, when the proportion of test-set is lower than or equal to 60%, SA behaves better than CN but worse than RA; when the proportion is higher than 60%, the performance of SA is the best.

(2) The performance of SA is still good even the proportion of the training set is relatively low, especially in precision. Probably because SA makes better use of the data indirectly related to the target network, and when the quantity of edges in the target network is small, its advantages become obvious.

Compared with CN and RA, the amount of calculation for SA is mainly concentrated in the extraction of subnetworks, which increases with the number and operation length of subnetworks. The lower the number and length, the closer to CN.

6 Discussion

In this paper, a method of link prediction on the basic of supernetwork and attention mechanism is proposed. This method provides a way of how to integrate different types of data for a specific target by relatively easily operation and less complex calculation than machine learning. The effectiveness of the model and its ability to extract the valuable information of the multi-source data are tested by the computational experiments. The performance of the method is better on the AUC index, which means that our method can extract useful information from multiple user related data to a certain extent. Meanwhile, this method may also be used in other fields, as long as there are additional attributes of the target networks.

This method can also be regarded as a kind of general system with two key ideas, information extraction and the combination of various relationships, and each idea can be further expanded. For information extraction, with the help of the operating unit, we can extract different kinds of subnetwork according to the demand. We can also study the upper limit of the operation length. For the next, distraction mechanism may be considered to improve efficiency.

The study of link prediction in supernetwork can assist us to further explore the multiple information in the era of information diversification, and for link prediction and supernetwork, there are still many interesting problems worthy of research, such as the prediction of the superedge, which means the relationship within more than two objects.

Acknowledgement. This work is supported by the National Natural Science Foundation of China under Grants 71573247, 91746106.

References

1. Lü, L.: Link Prediction. Higher Education Press, Beijing (2013)
2. Nagurney, A.: Supernetworks. In: Resende, M.G.C., Pardalos, P.M. (eds.) Handbook of Optimization in Telecommunications, pp. 1073–1119. Springer, Boston (2006). https://doi.org/10.1007/978-0-387-30165-5_37
3. Liu, Y., Li, Q., Tang, X., et al.: Superedge prediction: what opinions will be mined based on an opinion supernetwork model? Decis. Support Syst. **64**, 118–129 (2014)
4. Fu, M., Qu, H., Moges, D., et al.: Attention based collaborative filtering. Neurocomputing (2018)
5. Liben-Nowell, D., Kleinberg, J.: The link-prediction problem for social networks. J. Assoc. Inf. Sci. Technol. **58**(7), 1019–1031 (2007)
6. Adamic, L.A., Adar, E.: Friends and neighbors on the web. Soc. Netw. **25**(3), 211–230 (2003)
7. Zhou, T., Lü, L., Zhang, Y.C.: Predicting missing links via local information. Eur. Phys. J. B **71**(4), 623–630 (2009)
8. Hasan, M.A., Chaoji, V., Salem, S., et al.: Link prediction using supervised learning. In: SDM06: Workshop on Link Analysis, Counter-Terrorism and Security, pp. 798–805 (2006)
9. Wang, Z., Liang, J., Li, R., et al.: An approach to cold-start link prediction: establishing connections between non-topological and topological information. IEEE Trans. Knowl. Data Eng. **28**(11), 2857–2870 (2016)
10. Lin, D.: An information-theoretic definition of similarity. In: Fifteenth International Conference on Machine Learning, pp. 296–304. Morgan Kaufmann Publishers Inc. (1998)
11. Lorrain, F., White, H.C.: Structural equivalence of individuals in social networks. J. Math. Sociol. **1**(1), 67–98 (1971)
12. Jeh, G., Widom, J.: SimRank: a measure of structural-context similarity. In: Eighth ACM SIGKDD International Conference on Knowledge Discovery and Data Mining, pp. 538–543. ACM (2002)
13. Zhu, X., Tian, H., Cai, S., Huang, J., Zhou, T.: Predicting missing links via significant paths. EPL (Europhys. Lett.) **106**(1), 18008 (2014)
14. Zhu, X., Tian, H., Cai, S.: Predicting missing links via effective paths. Physica A **413**(11), 515–522 (2014)
15. Clauset, A., Moore, C., Newman, M.E.: Hierarchical structure and the prediction of missing links in networks. Nature **453**(7191), 98 (2008)
16. Anderson, C.J., Wasserman, S., Faust, K.: Building stochastic blockmodels. Soc. Netw. **14**(1), 137–161 (1992)
17. Zhang, M., Chen, Y.: Weisfeiler-Lehman neural machine for link prediction. In: ACM SIGKDD International Conference on Knowledge Discovery and Data Mining, pp. 575–583. ACM (2017)
18. Li, X., Chen, H.: Recommendation as link prediction in bipartite graphs: a graph kernel-based machine learning approach. Decis. Support Syst. **54**(2), 880–890 (2013)
19. Sheffi, Y.: Urban Transportation Networks: Equilibrium Analysis with Mathematical Programming Methods. Prentice-Hall, Englewood Cliffs (1984)
20. Nagurney, A., Dong, J.: Supernetworks: Decision-Making for the Information Age. Edward Elgar Publishing, Cheltenham (2002)

21. Estrada, E., Rodríguez-Velázquez, J.A.: Subgraph centrality and clustering in complex hyper-networks. Physica A Stat. Mech. Appl. **364**(C), 581–594 (2006)
22. Frank, H.P., Wooders, M., Kamat, S.: Networks and farsighted stability. J. Econ. Theory **120** (2), 257–269 (2005)
23. Wang, Z.: Reflection on supernetwork. J. Univ. Shanghai Sci. Technol. **33**(3), 229–237 (2011)
24. Liao, F., Arentze, T., Timmermans, H.: Multi-state supernetwork framework for the two-person joint travel problem. Transportation **40**(4), 813–826 (2013)
25. Liao, F.X.: Joint travel problem in space–time multi-state supernetworks. Transportation **4**, 1–25 (2017)
26. Nagurney, A., Toyasaki, F.: Supply chain supernetworks and environmental criteria. Transp. Res. Part D Transp. Environ. **8**(3), 185–213 (2003)
27. Yamada, T., Imai, K., Nakamura, T., et al.: A supply chain-transport supernetwork equilibrium model with the behaviour of freight carriers. Transp. Res. Part E Logistics Transp. Rev. **47**(6), 887–907 (2011)
28. Xi, Y., Dang, Y.: Method to analyze robustness of knowledge network based on weighted supernetwork model and its application. Syst. Eng. Theory Pract. **27**(4), 134–140 (2007)
29. Du, Y., Liu, X.: Research on key subject recognition method of knowledge-based super-network under target guidance. Sci. Technol. Prog. Policy **32**(23), 129–134 (2015)
30. Wang, N., Xu, W., Xu, Z., et al.: A survey on supernetwork research: theory and applications. In: Control Conference, pp. 1202–1206. IEEE (2016)
31. Liu, Y., Tang, X., Li, Q., et al.: Superlink prediction. Manag. Rev. **241**(2), 137–145 (2012)
32. Berlingerio, M., Coscia, M., Giannotti, F., et al.: Foundations of multidimensional network analysis. In: International Conference on Advances in Social Networks Analysis and Mining, pp. 485–489. IEEE (2011)
33. Eck, N.J.V., Waltman, L.: Software survey: VOSviewer, a computer program for bibliometric mapping. Scientometrics **84**(2), 523–538 (2010)

Dynamics of Deffuant Model in Activity-Driven Online Social Network

Jun Zhang[1,2(✉)], Haoxiang Xia[1(✉)], and Peng Li[2]

[1] Dalian University of Technology, Dalian 116024, Liaoning, China
Zhangjun2009tao@126.com, hxxia@dlut.edu.cn
[2] Shandong University of Technology, Zibo 255000, Shandong, China
sdutlp@163.com

Abstract. In many social system the interactions among the individuals are rapidly changing and are characterized with timing. The dynamics of social interaction constantly affects the development of their opinions. However, most of the opinion evolution models characterize interpersonal opinion in static, structural properties of the network such as degree, cluster and distance. In this paper, an Deffuant opinion model based on the activity-driven network is developed to examine how different activity distribution effects the dynamics of opinion evolution. When the activity distribution complies with power-law distribution or random distribution, phase transition transform from polarization to consensus when threshold is 0.6 and 0.4, respectively. In the process of opinion formation the distribution of opinion clusters' scales are complying with power-law distribution. Especially, under the power-law distribution the opinion disparity of the two clusters in polarization state is lower than the others, which means that the burst of the activity helps the individuals converging in opinion clusters in values. Finally we show that the speed to reach stable is influenced by the type of activity distribution. The simulation on power-distribution and random distribution need more time steps to get steady state.

Keywords: Activity-driven network · Temporal network · Opinion dynamic
Deffuant model

1 Introduction

More and more people interact with others by publishing posts or replying, which will facilitate the formation and evolution of public opinions online [1]. In particular, affordable and ubiquitous information and communication technologies (ICT) promote the online interactions changing rapidly over time [2, 3]. The underlying structure of the network and on the temporal activity patterns of humans, deeply influence the formation and the evolution of opinion on social media [4–6]. Consequently the relationships between the time-varying activities and the dynamics of opinion are worthy to deeply explore.

In general models about opinion dynamics consist of the mechanism of interaction and the network of interaction. They can be classified into two classes according to whether the variable that represents the opinion of an agent is discrete or continuous [7]. The voter model [8], the Galam majority rule model [9] and the Sznajd model [10] are

© Springer Nature Singapore Pte Ltd. 2018
J. Chen et al. (Eds.): KSS 2018, CCIS 949, pp. 215–224, 2018.
https://doi.org/10.1007/978-981-13-3149-7_16

the most common discrete opinion models. Deffuant model [11] and Hegselmann and Krause model [12] are the most typical continuous opinion models. In the above model, the opinion dynamics are always described by stationary states statistics, such as opinion clusters, phase transaction, the difference of opinion, etc. To the Deffuant model, the emphasis are explore the role of social influence [11], which is include but is not limited to threshold d and convergence u. For example, the value of d is proved to be used for control the scale of emerging clusters that will contribute to the phase transition of collective opinion on the complete networks, the square lattices, the random networks, and the scale free networks [13]. The mental capacity, the propaganda, and the fraction of conformists are important effect factors to the Deffuant models [14–16] In addition there are studies explore the effect of dynamical affinity or the local topology of the network surrounding the individuals in Deffuant model, which show interesting behaviors in regards to the structure of the social networks and their correlation with the opinion formation process or the opinion structures [17].

However most of the Deffuant models are evolved in static networks, which ignore the dynamic of social interaction network emerging from the timing of interacting. But various real-world social interaction systems, such as Twitter, Email messages, blogs broadcast etc., are characterized by processes whose timing and duration are defined on a very short time scale [18, 19]. Recent investigations about online social networks proved that the dynamics of user interactions may be more important than these static relationship networks in determining how information flows and opinion flows shuttle in a social network [20–22]. Specially, the burst nature of human interactions slows down the information spreading and results in the limited scope [23, 24]. Therefore, Maxi San Miguel has carried out exploration based on vote model on temporal network. They focused on the temporal variation in networks' connectivity patterns and the ongoing dynamic processes [25]. Andrzej study the influence of correlation between the activity of an individual and its connectivity on the process of opinion formation on social network [26]. These results represented that human activity plays a role in opinion formation, which is worthwhile to take it into account in the research of opinion model [27].

Motivated by the above analysis, in this paper, we study the opinion dynamics in Deffaunt model specifically devised for a class of time-varying networks, namely activity-driven networks. It is presented for describing the dynamics of a social network with individuals having activity potential and synthesis the interaction dynamics and topology dynamics together [28]. The topological structure of this kind of time-varying networks is measured by the distribution of activity of the nodes. Here we emphasis how activity distribution affect the evolution of opinion. We applied agent-based simulation to compare the effects of three activity distributions on the opinion dynamics of Deffuant model. We consider the normal distribution (corresponding to uniform probability of being active), the random distribution (being active disorder) and the power-law (to reproduce heterogeneous activity patterns) distributions. The main observation is that the phase transition, probability of the scales for opinion clusters, the opinion of each opinion clusters, and the speed to reach stable are influenced by activity distribution, called the structure of the time-varying network.

2 Deffuant Model on Activity-Driven Network

Social media is a typical communication platform, on which the relationships and the interaction among users build new channels for information diffusion and opinion formation. Thus, opinion dynamics is not only depending on online social networks where connections among nodes are long-lasting [29], but also on temporal network driven by individual's instant activity. That means on condition of the individual being active, the opinion interaction has possibility.

Here we illustrate our model with Fig. 1. In online communication platforms like Twitter, there is social network $G(N, E)$. $N = \{1, 2, \ldots, n\}$ is the set of nodes and their social relationships are tagged by the set $E = \{e(i,j)|i,j \in N\}$. Each node i has an activity potential $a_i = activity_i / \sum activity_i$, which is the probability per unit time to create new tweet or retweet online. $activity_i$ is the number of activities of node i over a period. Thus, a_i are bounded in the interval $[\varepsilon, 1]$ and comply with a given probability distribution $p(a)$. Additionally, node i has an continuous opinion x_i, which is bounded in the interval $[0, 1]$. For two nodes i and j, the opinion difference between them is defined as $\Delta x_{ij} = |x_i(t) - x_j(t)|$.

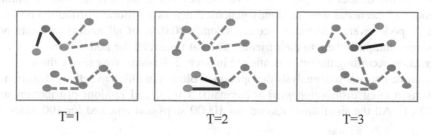

$$T=1 \qquad\qquad T=2 \qquad\qquad T=3$$

Fig. 1. The colored nodes and edges constructed the activity-driven network based on online social network. Red node is in the active state, who will adjust opinions according to the opinions hold by his neighbors on social network. Grey nodes are in the inactive state, who can't interact with their friends even if they are active. The structure of temporal networks is determined by (a), where a means the activity potential. For Twitter, $p(a)$ is power-law distribution and the slope value will be between -1.8 and -3. (Color figure online)

In previous work about Deffuant models, two agents meet and adjust their opinion when their difference of opinion them is smaller than threshold d. While in activity-driven network based on online social network, the opinion interaction will be constrained by both the social network and the activities. Generally at time t, only the nodes, who are in active state and receive the information from neighbors, can change opinion according the interaction rules. Hence, we assume an opinion interaction on activity-driven network according to the following rules:

- At each discrete time step t, there exists an online social network $G(N, E)$.
- With probability a_i node i becomes active. If node i is active, he can take one of the following behaviors randomly. One is creating a new message. The other is

forwarding a message received from node j, where exists an edge between i and j and $e(i,j) \in E$. Also the message is randomly chosen from his information storage. If node i is not active, it will do nothing. The forwarding behavior is recorded as information interaction.

- For two nodes, i and j, if there exists information interactions and $\Delta x_{ij} \leq d$, node i can update opinion as $x_i(t+1) = x_i(t) + \mu[x_j(t) - x_i(t)]$. If the opinion difference Δx_{ij} between them is larger than the tolerance threshold d, that means the opinion difference is out of his considerable range, so he keeps his current opinion.

3 Results

We analyze the effects of individuals' activity on opinion formation of Deffuant model by simulation experimental method. All the simulations based on the online social network $G(N, E)$, which is sampled from the Tencent microblog dataset [30]. $G(N, E)$ included records of 1509 node in two month, consisting of 10396 follower-followee relationships. The topology of it has small-world properties [20], where $<L> = 3.25$, $<C> = 0.4$, $M = 0.15$ and $<k> = 17.7$. The probability distribution of activities of the nodes in the dataset is $p(a) = a^{-2.7}$. At the beginning of each simulation, initially opinions of the nodes were randomly generated across a uniform distribution on $[0, 1]$ and 30 nodes were selected as seeds randomly (0.02% of all nodes) to create new messages and push them to their friends. Then at time t, all the nodes carry on opinion interaction according the rules mentioned in Sect. 2. Besides, we execute the model on the sample network and analysis the opinion evolution at different activity distribution, such as normal distribution $p(a) = \text{Normal}(0.5, 0.5)$, and random distribution with $a \in [0, 1]$. All the simulations carried out 10000 steps and repeated for 100 times.

3.1 The Opinion Phase Transition

The previous studies of Deffaut model showed that threshold values leading to 3 convergence station, consensus, polarization and fragmentation. When the threshold values are lower than 0.3, several opinion clusters can be observed and it is called fragmentation. When the threshold value is 0.3, there exits two opinion clusters and means polarization. When the threshold values are higher than 0.5, there exists only one opinion cluster, named consensus. Figure 2 shows the different results about relationships between the threshold values and the convergence station. One is no matter what the activity distributions are; there always exist some isolated nodes whose opinion cannot belong to any opinion cluster. The second is when the activity distribution is complied with power-law, the consensus will appear at $d = 0.6$. The effect of busty is lessening the level of consensus. In Fig. 2(b), polarization and consensus come up at $d = 0.2$ and 0.4 separately, which shows the uniform probability of being active is speed up the gathering of the opinion.

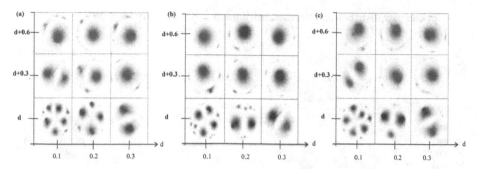

Fig. 2. The results of opinion phase transition with the increase of threshold d increased at different activity distribution. ($\mu = 0.5$, nodes with same color means they have same opinion, the) (a) $p(a) = a^{-2.7}$. Polarization is observed at $d = 0.3$. Consensus is observed at $d = 0.6$. (b) $p(a) = \text{Normal}(0.5, 0.5)$. Polarization is observed at $d = 0.2$. Consensus is observed at $d = 0.5$. (c) $p(a)$ is random distribution $a \in [0, 1]$. Polarization is observed at $d = 0.3$. Consensus is observed at $d = 0.5$.

3.2   The Probability of the Scales of the Opinion Clusters

In this study, we also concern how different activity distributions affect the scales of the opinion clusters. Figure 3 shows the results in details. The scale of opinion cluster is measured by N_{Ci}, where N_{ci} is the number of nodes in opinion cluster C_i. Most notably, the final macroscopic state is found to be difficult to reach complete consensus by ways of opinion detentions [26]. This is usually presented as the majorities of the individuals holding the same opinion but the rest holding many different opinions [31]. In Fig. 3 it can be seen that the scales of opinion clusters are complying with power-law distribution, except the results of random distribution. It means that the simulation results of the opinion model in this paper reproduce the distribution of the number of opinion clusters on social media by ways of activity driven.

3.3   The Difference of Clusters' Opinion

The difference of clusters' opinion reflects the degree of bias of opinion in the local network. In Fig. 4 we show the proportion of the scales and the opinion values for each opinion clusters at different convergence station and different activity distributions. Table 1 compares the statistics of the opinion and the scale of Top N opinion clusters in details.

In fragmentation state the variance of opinion among the Top4 opinion clusters are 0.28 for power-law distribution, 0.49 for normal distribution, and 0.58 for random distribution. All the scales of Top 4 opinion clusters are around 20% of the number of the whole network. At the same time the opinion of the cluster are well-distributed in [0, 1].

In polarization state the variance of the opinion between the Top2 opinion clusters are 0.23 for power-law distribution, 0.33 for normal distribution and 0.3 for random distribution. Meanwhile, the scales of the two largest opinion clusters in power-law

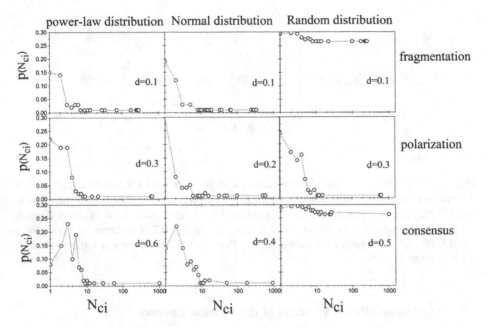

Fig. 3. The probabilities of the scales for opinion clusters at different convergence station and different activity distributions, which are the statistical results of the 100 simulations ($\mu = 0.5$). The left column is for power-law distribution $p(a) = a^{-2.7}$, middle column for normal distribution $p(a) = \text{Normal}(0.5, 0.5)$ and right column for random distribution $a \in [0, 1]$. Top row contains plots for fragmentation state, second row shows the plots for polarization station and third row shows the plots for consensus station.

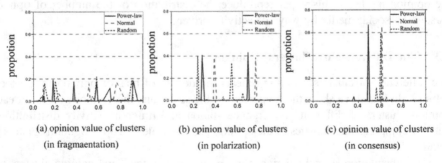

Fig. 4. The proportion of the scales and the opinion of each opinion clusters at different convergence station and different activity distributions, which are the statistical results of the 100 simulations ($\mu = 0.5$). The black line is for power-law distribution $p(a) = a^{-2.7}$. The red line is for normal distribution $p(a) = \text{Normal}(0.5, 0.5)$. The blue line is for random distribution $a \in [0, 1]$. (Color figure online)

Table 1. The proportion and the opinion value of the Top N opinion clusters

State	Power-law distribution	Normal distribution	Random distribution
Fragmentation d = 0.1	N_{C1} = 286, opinion = 0.19 N_{C2} = 271, opinion = 0.37 N_{C3} = 241, opinion = 0.71 N_{C4} = 271, opinion = 0.92	N_{C1} = 241, opinion = 0.05 N_{C2} = 181, opinion = 0.12 N_{C3} = 256, opinion = 0.51 N_{C4} = 271 opinion = 0.77	N_{C1} = 256, opinion = 0.06 N_{C2} = 271, opinion = 0.21 N_{C3} = 316, opinion = 0.59 N_{C4} = 286, opinion = 0.92
Polarization d = 0.2/0.3	N_{C1} = 650, opinion = 0.34 N_{C2} = 614, opinion = 0.66	N_{C1} = 599, opinion = 0.26 N_{C2} = 505, opinion = 0.73	N_{C1} = 623, opinion = 0.28 N_{C2} = 587, opinion = 0.7
Consensus d = 0.4/0.5/0.6	N_{C1} = 1011, opinion = 0.5	N_{C1} = 903, opinion = 0.52	N_{C1} = 983, opinion = 0.51

distribution are bigger than the other two distributions. Although the simulation with normal distribution transferred to polarization state at threshold d = 0.2, but the conflict of the two opinion clusters is more violent. In contrast, random activity will reduce the conflict among individuals.

Comparing the results in consensus state, the power-law distribution shows the most conformance. Because the scale of the largest opinion cluster is 1011 and its' opinion value is 0.5. As mentioned in [32], in the process of becoming consensus the system should experience opinion bifurcation, which means small clusters with different opinion values combined into a union gradually. In power-law condition, the difference of the values between opinion clusters is smaller than those of the two distributions. Thus the bursty of the activity helps the individuals converging in opinion clusters. However, the uniform probability of being active, named normal distribution is not benefit to the convergence of the opinion clusters' value.

3.4   The Difference of Nodes' Opinion in the Whole Network

The difference of nodes' opinion in the whole network reflects the overall degree of bias of opinion in the whole network. To measure how individual opinions differ at each step, we here use the standard deviation of the individuals' opinions:

$$s_{opinion}(t) = \sqrt{\frac{1}{N}\sum_{i=1}^{N} x_i(t) - \overline{X(t)}} \tag{1}$$

Where $\overline{X(t)}$ is the mean value of all individuals' opinion at time t.

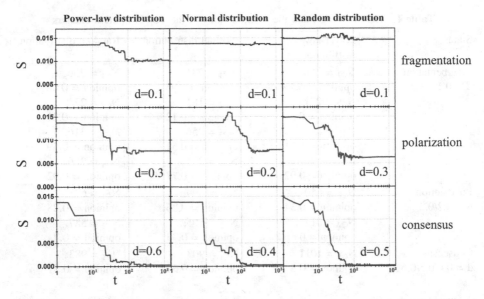

Fig. 5. The difference of opinion in the whole network. $s_{opinion}$ varies with time t for different activity distributions, which is the average results of 100 simulations. The left column is for power-law distribution $p(a) = a^{-2.7}$, middle column for normal distribution $p(a) =$ Normal$(0.5, 0.5)$ and right column for random distribution $a \in [0, 1]$. Top row contains plots for fragmentation state, second row shows the plots for polarization station and third row shows the plots for consensus station. $s_{opinion}$.

Figure 5 shows $s_{opinion}$ varying with time t for different activity distributions. Overall when the threshold d increased the $s_{opinion}$ decreased. Since the threshold d expressed the tolerance of nodes' to his neighbors, the more the tolerance is, the more the convergence are.

With the increase of time t, the difference of opinion will decrease and reaches a certain value, which presents the individuals' opinion reach a steady. Under the condition of the random distribution, the $s_{opinion}$ achieved stable at $t = 150$. To the power-law distribution and normal distribution, the $s_{opinion}$ achieved stable at $t = 200$, which is similar with the result in [33]. The results can be illustrated as follows. Although the individuals with high activity contact others more frequently, most of their contacts have low activity rate [34]. Because of the individuals having low activity with a small probability, the total number of interacting individuals would be reduced. So in the experiment on power-law distribution, more time is needed to reach stability.

4 Conclusion

Network model plays an important role to many complex systems. Because of the function to describe the rapidly changing interactions among individuals, the activity driven network is gain more and more interest. In this paper, we have studied the opinion formation of Deffuant model in online social network, named Tengxun Weibo. In our model, we have taken into account the activity of individuals, on which the activity-driven network is generated dynamically. Opinion interaction happens when individuals are active. According to the simulation results, we have found that the activity distribution, namely structure of activity-driven network, affects the opinion dynamic. Because the busrty of activity, the power-law distribution will decrease the speed to get steady. Yet it help individuals' opinion become similar for the smallest variance of opinion among the Top N opinion clusters and $s_{opinion}$. Under the condition of the normal distribution, the simulation transferred to polarization state at threshold $d = 0.2$ and the conflict of the two opinion clusters is more violent. It means that uniform probability of being active will increase the chance to communicate, but the threshold d inhibition the interaction between the two having large opinion gap.

Funding Statement. This work was supported in part by the National Natural Science Foundation (grant numbers 71401024, 71371040 and 71801145).

Conflicts of Interest. The author(s) declare(s) that there is no conflict of interest regarding the publication of this paper.

References

1. Le, T.N., Wu, P., Chan, W., et al.: Predicting collective sentiment dynamics from time-series social media. In: International Workshop on Issues of Sentiment Discovery and Opinion Mining, pp. 1–8. ACM (2012)
2. Sobkowicz, P., Kaschesky, M., Bouchard, G.: Opinion mining in social media: modeling, simulating, and forecasting political opinions in the web. Gov. Inf. Q. **29**(4), 470–479 (2012)
3. Tsytsarau, M., Palpanas, T., Castellanos, M.: Dynamics of news events and social media reaction. In: ACM SIGKDD International Conference on Knowledge Discovery and Data Mining. ACM, pp. 901–910 (2014)
4. Barabási, A.: The origin of bursts and heavy tails in human dynamics. Nature **435**(7039), 207 (2005)
5. Iribarren, J.L., Moro, E.: Impact of human activity patterns on the dynamics of information diffusion. Phys. Rev. Lett. **103**(3), 038702 (2009)
6. Malmgren, R.D., Stouffer, D.B., Motter, A.E., et al.: A poissonian explanation for heavy tails in e-mail communication. Proc. Natl. Acad. Sci. U. S. A. **105**(47), 18153–18158 (2008)
7. Shang, Y.: Consenseus formation of two-level opinion dynamics. Acta Math. Sci. (Engl. Ser.) **34**(4), 1029–1040 (2014)
8. Axelrod, R.: The dissemination of culture: a model with local convergence and global polarization. J. Confl. Resolut. **41**(2), 203–226 (1997)
9. Thompson, R.: Radicalization and the use of social media. J. Strat. Secur. **4**(4), 167–190 (2011)
10. Nickerson, R.S.: Confirmation bias: a ubiquitous phenomenon in many guises. Rev. Gen. Psychol. **2**(2), 175–220 (1998)
11. Deffuant, G, Neau, D, Amblard, F.: Mixing beliefs among interacting agents. Adv. Complex Syst. **03**(01n04), 0000007 (2000)
12. Hegselmann, R., Krause, U.: Opinion dynamics and bounded confidence models. Anal. Simul. J. Artif. Soc. Soc. Simul. **5**(2), 2 (2002)
13. Lorenz, J., Urbig, D.: About the power to enforce and prevent consensus by manipulating communication rules. Adv. Complex Syst. **10**(02), 251–269 (2007)

14. Weisbuch, G.: From anti-conformism to extremism. J. Artif. Soc. Soc. Simul. **18** (2015)
15. Malarz, K., Gronek, P., Kulakowski, K.: Zaller-Deffuant model of mass opinion. J. Artif. Soc. Soc. Simul. **14**(1), 2 (2011)
16. Timothy, J.J.: How does propaganda influence the opinion dynamics of a population? Int. J. Mod. Phys. C **27**(05) (2016)
17. Gargiulo, F.: Opinion dynamics in a group-based society. Eur. Phys. Lett. **91**(5), 2067–2076 (2010)
18. Holme, P., Saramäki, J.: Temporal networks. Phys. Rep. **519**(3), 97–125 (2012)
19. Ghoshal, G., Holme, P.: Attractiveness and activity in Internet communities. Phys. Stat. Mech. Appl. **364**, 603–609 (2005)
20. Kossinets, G., Kleinberg, J., Watts, D.: The structure of information pathways in a social communication network. In: Proceeding of 14th ACM International Conference on Knowledge Discovery and Data Mining, pp. 435–443. ACM (2008)
21. Viswanath, B., Mislove, A., Cha, M., Gummadi, K.P.: On the evolution of user interaction in Facebook. In: Proceedings of the 2nd ACM Workshop on Online Social Networks, pp. 37–42. ACM (2009)
22. Wilson, C., Boe, B., Sala, A., Puttaswamy, K.P., Zhao, B.Y.: User interactions in social networks and their implications. In: Proceedings of 4th ACM European Conference on Computer Systems, pp. 205–218. ACM (2009)
23. Bakshy, E., Hofman, J.M., Mason, W.A., et al.: Everyone's an influencer: quantifying influence on twitter. In: ACM International Conference on Web Search and Data Mining, pp. 65–74. ACM (2004)
24. Guille, A., Hacid, H., Favre, C., et al.: Information diffusion in online social networks: a survey. ACM Sigmod Rec. **42**(2), 17–28 (2013)
25. Fernández-Gracia, J., Eguíluz, V.M., Miguel, M.S.: Timing interactions in social simulations: the voter model. In: Holme, P., Saramäki, J. (eds.) Temporal Networks. Understanding Complex Systems, pp. 331–352. Springer, Heidelberg (2013). https://doi.org/10.1007/978-3-642-36461-7_17
26. Grabowski, A.: Opinion formation in a social network: the role of human activity. Phys. Stat. Mech. Appl. **388**(6), 961–966 (2012)
27. Patterson, S., Bamieh, B.: Interaction-driven opinion dynamics in online social networks. In: The Workshop on Social Media Analytics, pp. 98–105. ACM (2010)
28. Perra, N., Gonçalves, B., Pastorsatorras, R., et al.: Activity driven modeling of time varying networks. Sci. Rep. **2**(6), 469 (2012)
29. Vespignani, A.: Evolution and Structure of the Internet. Cambridge University Press, Cambridge (2004)
30. Jalali, Z.S., Rezvanian, A., Meybodi, M.R.: Social network sampling using spanning trees. Int. J. Mod. Phys. C **27**(05) (2016)
31. Sobkowicz, P., Sobkowicz, A.: Dynamics of hate based Internet user networks. Eur. Phys. J. B **73**(4), 633–643 (2010)
32. Kurmyshev, E., Juárez, H.A., González-Silva, R.A.: Dynamics of bounded confidence opinion in heterogeneous social networks: concord against partial antagonism. Phys. Stat. Mech. Appl. **390**(16), 2945–2955 (2011)
33. Li, D., Han, D., Ma, J., et al.: Opinion dynamics in activity-driven networks. EPL **120**(2), 28002 (2017)
34. Guo, Q., Lei, Y., Jiang, X., et al.: Epidemic spreading with activity-driven awareness diffusion on multiplex network. Chaos Interdisc. J. Nonlinear Sci. **26**(4), 3200 (2016)

Identifying Factors that Impact on the Learning Process of Sewing Workers on an Assembly Line

Thanh Quynh Song Le[✉] and Van Nam Huynh

Japan Advanced Institute of Science and Technology, Nomi, Japan
lstquynh@jaist.ac.jp

Abstract. Today, the economic environment offers many opportunities due to the Open-Economy Policy. Most apparel companies are engaged in mass customization production and are working towards shorter product cycle times and production runs. A better understanding of the learning process of sewing workers will allow the clothing industry to improve productivity in its manufacturing process. This article indicates the factors that have an effect on the learning process of sewing employees and examines the statistical significance of test results based on empirical data. The results indicate that three factors significantly affect the learning rates of sewing workers: previous experience, the structure of the task, and job complexity. Specific knowledge about these factors could form a basis for developing worker assignment methods, the calibration of cross-training programs, and determining productivity.

Keywords: Learning of worker · Learning rate · Worker assignment

1 Introduction

Nowadays, workers in a manufacturing environment must continually learn many new skills, technologies and processes in order to keep up with the move toward shorter product cycle times and production runs. Consequently, the learning process of workers is becoming an increasingly important factor in improving manufacturing productivity. Understanding the process of learning is also crucial for a broad range of production research, including production scheduling, setting standard time, estimating labor costs, and optimizing cross-training programs for workers. Considering the learning process of workers will help to improve the accuracy of production planning.

The learning process of workers has received a considerable amount of attention from researchers. The existence of learning effects in manufacturing processes has been widely confirmed and accepted from many previous studies, such as those by Conway and Schultzto, 1959; Cochran, 1960; Day and Montgomery, 1983; and Venezia, 1985 [1]. In 1966, Keachie and Fontana [2] were the first researchers to consider learning effects when calculating optimal lot-sizes. They concluded that if learning effects occur during the production of the first lot, the lot will be finished earlier than when there are no learning effects. Kumar and Goswami [3] researched the learning effect of the unit production time on optimal lot sizes in an uncertain environment process.

© Springer Nature Singapore Pte Ltd. 2018
J. Chen et al. (Eds.): KSS 2018, CCIS 949, pp. 225–236, 2018.
https://doi.org/10.1007/978-981-13-3149-7_17

Edmondson et al. [4] studied the learning process in small groups to present a variety of terms, concepts, and methods. The results show that teamwork played a crucial role in organizational learning and that longstanding interest made effective organizational work teams. Letmathe et al. [5] researched how the different forms of knowledge transfer affected the process of learning and the task performance of individuals. They concluded that explicit knowledge transfer was most beneficial with respect to manufacturing performance in quality products and production time when a new task was applied. In the study of Grosse and Glock [6], they investigated the effect of learning on the performance of an order picker. They collected the empirical data in an order picking warehouse and fitted this data with the different learning curve models. The results show that learning rates of order pickers varied between 84.25% and 97.67%. In 2014, Grosse and Glock [7] continuously researched an approach to model worker learning in order picking. In particular, the results show that the learning process of worker should be considered when planning the order picking task for the better prediction of order throughput time.

In a sewing manufacturing process, workers are regularly assigned or reassigned to jobs based on their performance rating. The performance of a sewing worker is measured in two ways: by the worker's qualification and by their skills [8]. Qualification represents a worker's ability to operate certain types of machines and/or to perform certain types of manual work. A higher qualification shows the worker's ability to carry out more difficult jobs. The workers' skills are graded according to the level of skill mastery in performing those jobs. A higher level of skill mastery results in a shorter amount of time taken or greater efficiency in doing the work. Qualification grading is usually assessed according to national or industry standards [9]. On the other hand, skill grading is job-specific. Furthermore, skill grading is typically set and used by the company for rating the workers' performance. The most popular performance rating system is the one by Westinghouse [10], as it enables production managers to predict operator performance and therefore production capacity accurately. Also, the Westinghouse system allows managers to align operator performance with production unit productivity and achieve their quality goals.

However, the sewing manufacturing environment has changed drastically in response to the Open-Economy Policy. Now, most clothing production firms are engaged in mass customization production. Product designs, customer quality requirements, materials, and even the equipment used to make the clothing change at a rapid pace. The size of orders are getting smaller and smaller. In addition, customers today are more demanding than ever before: the product quality must be better, the cost must be lower, and delivery time is non-negotiable with a large penalty imposed for any delay. In this new environment, the sewing workers must learn new tasks far more often.

The traditional methods for assigning workers to tasks have failed to accurately predict the workers' performance needs for planning the work, and have also failed to encourage the workers to develop the new skills and skill mastery needed for improved quality and productivity. The traditional methods failed because do not factor in the need for workers to spend a period of time learning and becoming familiar with new tasks before they reach a steady rate of productivity. Lacking a thorough understanding of the individual workers' learning characteristics, the employees' task assignments were founded on elements that did not necessarily relate to productivity, let alone productivity under the condition of continuous changes.

Since sewing employees respond differently to changes, particularly with respect to learning new skills, worker assignments based on the workers' individual learning personalities has a significant potential to improve productivity in a production process [11]. That is, how a sewing worker is expected to respond to change becomes as significant, if not more so, than how efficiently the worker is performing a current assignment. This is the reason why the assignment of an individual worker to a task based on their learning behavior characteristics has received relatively more attention in the literature.

On the other hand, the learning process of a worker on an assembly line is influenced by many other factors associated with manufacturing, such as the amount of training time given to the workers, the training methods used, the machines the workers uses for production, design changes, and process setting. Identifying the factors that impact the learning process of sewing workers will be useful for predicting the workers' performance and optimizing the assignment of workers to tasks based on each sewing worker's learning characteristics and characteristics of task.

This paper aims to address the factors that affect the learning process of sewing workers and is based on empirical data. Knowledge about the effects of these factors on individual learning will be a premise for predicting the performance ratings of sewing workers and help supervisors to assign sewing workers to suitable tasks.

The structure of this paper is organized as follows. In Sect. 2 we identify the factors that have an effect on the learning process of sewing workers using the Delphi method. Consequently, in Sect. 3, we apply a methodology to a case study to test for significant effects due to these factors on the learning rate of sewing workers. Finally, Sect. 4 summarizes the principal results by emphasizing the importance of this method and offers suggestions for future research.

2 The Method for Determining Factors that Influence the Learning Process of Sewing Workers on an Assembly Line

The learning process of sewing workers can include different individual parameters such as learning rate, steady-state productivity rate, or forgetting rate [12]. In this study, the measure of the individual sewing worker's learning process is defined based on the learning rate concept. The learning rate shows the difference in performing a task as a function of time. We represent the learning behavior of each worker by using a mathematical formula that was suggested in a model of the learning curve by Theodore Paul Wright [13]; the formula is

$$Learning\ rate = \frac{The\ time\ taken\ to\ produce\ the\ first\ 2N\ unit}{The\ time\ taken\ to\ produce\ the\ first\ N\ unit} \quad (1)$$

where N is the number of products that the worker had finished. The higher the learning rate is of a worker in a production process, the less difference there will be in the time taken to produce the first $2N$ and N products.

We will first determine all the factors that impact the learning rate of sewing workers based on information in our survey of the literature and consulting expert advice using the Delphi method. In essence, this method uses the experience and knowledge of experts, thus exploiting their opinions to identify and consider a problem and to find the optimal solution to the problem. The Delphi method was developed by the Rand Corporation in the early 1960s in response to problems in forecasting future events associated with the military potential of future technology and potential political issues and their resolution [14, 15]. The Delphi method is based on the following suppositions:

- Experts can reach a consensus about the questions in their field of expertise; the answers reached are likely to be correct; and the result is better than that obtained from a single expert.
- The personality dominance that could interfere with the independent judgment of individual experts in face-to-face interactions have to be eliminated, and therefore anonymity is required in the sense that no one knew who else was participating.

In order to use the Delphi method effectively, some requirements should be noted:

- It is important to select the right, qualified experts, who have experience in the field of the research, are honest, and are objective in their assessments and evaluations.
- It is also important to select the appropriate experts to consult regarding the issues to be investigated.

The Delphi method is also the subject of many criticisms. However, it is the only method that supports a group making a decision based on a group consensus. Today, the Delphi method is increasingly applied in complex studies in which a consensus must be reached. Meanwhile, its applications are widely spread. The procedure to determine the factors that influence the learning rate of sewing workers on an assembly line uses three main steps that are presented below.

2.1 Preparation

First, we identified experts who have both experience and knowledge about training, coaching and predicting the performance of sewing workers in companies. The success of the Delphi method depends on the precise selection of the experts on the panel. During the preparation process, we identified a group of nine experts. These experts are a group of experienced supervisors and group leaders who have been working at the company for a long time. Four experts are in the manufacturing department, three are in the planning department, and two are in the training department. These leaders also are very experienced sewing workers on assembly lines themselves. As leaders, these people know all the workers very well; they are now directly involved in managing, coaching and training the current and new sewing workers. The experts also have a deep understanding of the workers' performance. In addition to that, the experts showed good observation skills and all the experts are comfortable expressing their opinions.

We explained some benefits of the experts becoming a member of the Delphi panel, such as: (1) the opportunity to study, and develop their experience and knowledge from the consensus survey; and (2) enhancing their visibility both within their organization and outside it. These benefits offered the strong inducements needed to attract experts to participate in this part of our research.

Second, we developed a list of factors that could affect the productivity and learning process of the workers after reviewing the literature and previous studies. In addition, the definition of all supporting evidence/information necessary.

2.2   Using the Delphi Method to Determine Efficient Factors Process

First, the nine experts verified the elements or categories based on a collection of all the supporting evidence/information about the representative learning process of workers from the list of factors prepared in Step 1. In addition, we asked the experts to provide more criteria that they believe influences the learning process of workers.

Although the Delphi method suggests that the name of the experts and their positions should be hidden to respect their anonymity, in this case the experts welcomed having a general meeting to discuss the previous round and results. We therefore held a Delphi conference to reach a consensus on the relative importance levels of each factor that has an impact on the learning process of workers and on relative difference factors with sub-factors below. During this process, the experts also gave reasons for their judgments, suggested additions, take-aways and/or modifications to the information/evidence provided in order to reach a consensus. After that, the experts determined the final factors that have an effect on the learning process of workers.

Based on the general agreement in the answers from the experts participating in the Delphi panel questionnaires and the general Delphi experts' meeting, the learning process of sewing workers as represented by the learning rate concept will be influenced by three factors: two factors that are part of the manufacturing process, and one factor that a characteristic of the workers. These factors are shown in Fig. 1.

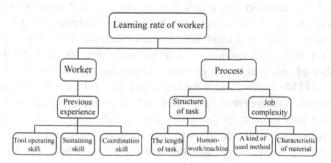

Fig. 1. Factors influencing the learning process of sewing workers on an assembly line.

The Factors from the Manufacturing Process that Influence the Sewing Workers' Learning Process

These two factors, namely job complexity and the structure of task, are aspects of the manufacturing process and directly influence the sewing workers' learning process.

Job Complexity

According to previous research, the content of a job will directly influence both productivity and the error rate of an assembly line [16, 17]. Furthermore, it is often reasonable to assume that job complexity has an effect on the learning process of sewing workers. The greater amount of information that is needed to complete a job successfully, the longer it would take to learn the task completely. If less information is needed, it would be easier to learn the task [18, 19]. In sewing manufacturing processes, job complexity consists of two sub-factors, such as a kind of used method and characteristic of material. In the mass customization production, the fabric/material being sewn is a factor that changes from one order to another, and the properties of the fabric/material will greatly affect both the standard time needed to complete the work successfully and the learning process of worker. For example, in the apparel industry, a fabric that wrinkles and/or stretches easily will make it more difficult for workers to finish the product. The workers must spend time becoming familiar with the fabric and learning to sew it properly, and therefore the learning process of the workers will take longer. It will also take longer for the workers to reach a consistent rate of productivity. If the fabric wrinkles easily and stretches easily, there will also be a lower level of productivity because the workers will need to make many careful adjustments during the sewing process.

The Structure of Task

In addition to job complexity, the structure of task always has an effect on the learning process of workers. On an assembly line, other workstations may have the same job complexity, such as the characteristics of a fabric and/or material, but the difference in the structure of task, including the length of time needed to successfully complete the task and the interactions between operator physical characteristics and workstation that affect operator performance (human-work/machine). The longer it takes to successfully complete a task, the longer it will take a person to learn to do that task. In this case, the value of the performance time of the first cycle will also be larger.

We should also consider that a task may contain several repetitions of sub-tasks. A sub-task is defined as "…a distinct describable and measurable subdivision of a task and may consist of one or several fundamental motions" [20]. We can extend this argument to say that all duties in the manufacturing process are a combination of a limited number of essential micro-motion components defined as the Method-Time Measurement (MTM) system, and therefore a task would be the sum of unique micro-motion components that appear in the overall task. We should include the sub-tasks that are part of each task in order to determine the overall structure of the task, and then compare the differences between the structure of a new task and another task, as the basis for estimating the workers' learning rate for the new task. For example, in the production of a shirt, two sleeves could be produced for each completed shirt so that the learning process for producing the sleeves would occur faster than for the shirt itself.

Furthermore, McCampbell and McQueen [21] suggested that a completely automatic operation such as an automatic screw machine would have a 100% learning rate, but a complex handmade task might be as low as a 60% learning rate. From this previous research, it is clear that if a job uses more automated processes to complete a product, the workers will learn to complete the product faster than when they must use a higher percentage of nonautomated processes.

The Factors that Are Characteristic of Sewing Workers

On an assembly line, sewing workers are the primary element that determines the productivity of the production process. Natural differences in the learning process occur between individuals who have previous experience. The previous experience of workers is a kind of tacit knowledge in the manufacturing process. It is very challenging to estimate how much "experience" we can say that a worker has. In a manufacturing environment, a sewing worker's previous experience is usually estimated according to the skill level of the worker. Crossman [22] says that skill consists mainly of selecting the correct method for each situation, and claims that the transfer of skill from one task to another takes place when the methods used for the old task are also suitable for the new one. However, the amount of skill transferred will depend more on which criteria are selected than on the mere coincidence of methods used in each task.

The previous experience of a sewing worker could also be determined by assessing the skill elements of a worker. Skill elements are the factors demonstrated through the operational behaviors of an operator or through interactions of an operator and workstation characteristics that effect the operator's performance of their tasks.

The group of experts verified a list of operational skill elements earlier identified by facilitators through an intensive search of the literature. Three aspects of operational skill elements were identified, as discussed below.

- Coordination skill: This skill demonstrates the sewing worker's ability to coordinate their thoughts and hands as they work.
- Sustaining skill: This is the sewing worker's ability to repeat operations consistently.
- Tool operating skill: The demonstrates the sewing worker's ability to use tools and assembly parts, including the interactions between the worker's physical characteristics and the workstation that affect the worker's performance.

2.3   Statistical Validation of the Effects of These Elements on the Learning Process of Sewing Workers

In this step, the actual data was collected to verify the effects of the three factors mentioned above on the learning process of sewing workers. The hierarchical multiple regression method is applied to answer questions [23, 24], such as (1) how well a set of three independent variables is able to predict the learning process of a sewing worker, and (2) which independent variable is the best predictor of the learning process of a worker.

3 Case Study

3.1 Participants and Measures

The current study took place at a vest assembly line where the manufacturing process has a lot of sewing tasks that incorporate worker-paced machinery, and workers are the critical factor that determines both productivity and the quality of the products produced.

The actual data had been collected from the assembly sewing line to test three factors. On the assembly line, there are 106 workstations; each workstation includes a worker and a machine. Regarding the previous experience factor, we base our assessment on the difference of the assigned skill levels of the workers. The skill level is separated into four categories of workers with skill levels 1, 2, 3 and 4. The various structures of the tasks involves the difference in the standard time needed to complete the task, the type of equipment employed, and the amount of automation used in completing the task. In addition, the measure of the job complexity in this study is based on the difference in the shape and length of seams, the method for locating a semi-finished product, the size of the semi-finished product, and the characteristics of the material used, such as elasticity, density, and the type of fibers in the fabric. The different levels of job complexity are shown in Table 1.

Table 1. Classification of job complexity levels

Classification criterion	Level "Easy"	Level "Average"	Level "Difficult"
The shape of the seam	Straight line	Curve	Circle
The length of the seam	Short	Average	Long
Method for locating a semi-finished product	Do not need to locate	1–2 points	3–4 points
Size of a semi-finished product	Small	Average	Large
The kind of material	Thick, flat and well-shaped	Thin and well- shaped	Thin, curly hem and badly shaped

3.2 Multiple Regression Analyses

Hierarchical multiple regression analysis is applied to estimate the relative contribution of the three variables to the prediction of the learning process of a sewingworker. In this model, to control for the incremental influence of previous experience, the structure of the task, and job complexity, three hierarchical regression steps are established. In these steps, the learning rate of a sewing worker is the dependent variable, and three independent variables will be used in the regression model in the following order: previous experience, the structure of the task, and job complexity.

In the model, the previous experience variables are coded this way: skill level 1 is coded as 1, skill level 2 is coded as 2, and skill levels 3 and 4 are coded as 3 and 4.

Similarly, the structure of a task is estimated as the standard time needed to complete the task. The different job complexities are coded with the easy level coded as 1, the average level is coded as 2, and the difficult level is coded as 3.

3.3   Result and Discussion

Correlations Among the Measures

The descriptive statistics, including the mean, standard deviations and Pearson's correlation coefficients between all the measures and variables are shown in Table 2.

Table 2. Descriptive statistics and Pearson's correlation coefficients

	Mean	Standard deviation	1	2	3	4
1. Learning rate	85.16	4.42	–	0.27	−0.38	−0.32
2. Previous experience	2.15	0.92	0.27	–	−0.07	0.54
3. Structure of the task	28.15	8.95	−0.38	−0.07	–	−0.14
4. Job complexity	1.90	0.78	−0.32	0.54	−0.14	–

Pearson's correlation coefficient measures the linear correlation between two variables. The values of Pearson's correlation will be given from +1 to −1. The results indicate that previous experience is positively correlated with the learning rate of a worker ($r = 0.27$). In sewing assembly line, when a worker has more previous experience with the current task, the value of the performance time of the first cycle will also be smaller; he/she doesn't need more time to learn and adapt with characteristic of task, and the difference between the time taken to produce the first 2N and N products will also be small. This means that when a worker has more experience with the current task that the worker will achieve a steady-state of productivity faster.

On the other hand, the structure of task and job complexity are negatively correlated with learning rate. The result is similar to that described in the literature and suggests that the learning rate of a sewing worker will decrease when he or she works on more complex tasks and structures of tasks. A complex process offers more opportunity than does a simple process to improve work methods and the performance time of workers. In this case, the value of the performance time of the first cycle will be larger, the workers need more time to learn, give more opportunity to make reductions in the required hours per unit.

Multiple Regression Analyses

Results from the statistical hypothesis testing in this current study all have P-values < 0.05; this means that all three independent variables including previous experience, structure of the task, and job complexity that have an effect on the learning rate of sewing worker. This conclusion is significant and supports the Delphi method's results (Table 3).

Moreover, when model 1 has only one independent variable, such as previous experience, we find that previous experience accounted for 6.5% of the total variance of

Table 3. Hierarchical multiple regression analysis

Model		Standardized coefficients	t	P-values	Adjusted R^2	ΔR^2
1	Previous experience	0.272	2.884	0.005	0.065	–
2	Previous experience	0.247	2.803	0.006	0.189	0.124
	Structure of the task	−0.363	−4.117	0.000		
3	Previous experience	0.626	8.072	0.000	0.551	0.362
	Structure of the task	−0.439	−6.642	0.000		
	Job complexity	−0.716	−9.151	0.000		

the learning process of a sewing worker ($R^2 = 0.065$), which is quite small. In step 2, when the structure of the task is entered into the model, the effect is 18.9%, an increase of 12.4% when compared with model 1. Finally, when the regression model includes all three variables, the model accounts for 55.1% of the total variance of the learning process. From this result, we can see that the learning process of a sewing worker should be estimated using the combination of and interaction between all three factors. In addition, job complexity has the highest standardized coefficient ($\beta = -0.716$), so it has the biggest effect on the variance of learning rate. The next largest impact is previous experience ($\beta = 0.626$), and the smallest impact is the structure of the task ($\beta = -0.439$).

4   Conclusion

In this .paper, we presented three factors have an effect on the learning process of sewing workers and verified the influence of the three factors by using the hierarchical multiple regression test. The results indicate that previous experience, the structure of the task, and job complexity significantly affect the variance of the learning rate of sewing workers. In this study, we have found that the factor that contributed the most to the variance of the learning process of sewing workers was job complexity. The next largest impact is previous experience, and the smallest impact is the structure of the task. Specific knowledge about these effects could aid in worker assignments, calibration of cross-training programs, and estimates of productivity.

The learning rate of a worker will be affected by the worker's previous experience a worker who has more experience doing the current task will achieve steady-state productivity faster. However, the amount of experience transfer will depend more on which criteria are selected than on the mere coincidence of methods between the tasks. The issue is even more complicated when an operator achieves excellent skills doing one type of work but is then assigned to a workstation with a different type of job. In this situation, the estimation of previous experience based on three sub-elements, such as coordination skill, sustaining skill, and tool operating skill must include multiple quantitative and qualitative factors that include uncertainty and precision should be applied in the manufacturing process. Moreover, in the future, the development of a systematic examination of the concept of job complexity through two semi-factors including the kind of method used and characteristics of the material is needed.

However, the criteria for classifying the degree of job complexity will depend on the particular characteristics of the type of work being evaluated.

In addition, in previous research, Wright (1936) reported that the learning rate of a worker was a constant number, an 80% learning rate. Many researchers who have investigated the application of learning curves to industrial engineering problems tend to assume their learning curves will follow the 80% learning rate. With the calculation of learning rate from actual data based on the difference of the three factors used in this study, there is enough evidence to support the assertion that learning rates vary depending on the products and across organizations.

References

1. Biskup, D.: A state-of-the-art review on scheduling with learning effects. Eur. J. Oper. Res. **188**(2), 315–329 (2008). https://doi.org/10.1016/j.ejor.2007.05.040
2. Keachie, E.C., Fontana, R.J.: Effects of learning on optimal lot size. Manag. Sci. **13**(2), B-102 (1966). https://doi.org/10.1287/mnsc.13.2.b102
3. Kumar, R.S., Goswami, A.: EPQ model with learning consideration, imperfect production and partial backlogging in fuzzy random environment. Int. J. Syst. Sci. **46**(8), 1486–1497 (2015). https://doi.org/10.1080/00207721.2013.823527
4. Edmondson, A.C., Dillon, J.R., Roloff, K.S.: 6 three perspectives on team learning: outcome improvement, task Mastery, and group process. Acad. Manag. Ann. **1**(1), 269–314 (2007). https://doi.org/10.1080/078559811
5. Letmathe, P., Schweitzer, M., Zielinski, M.: How to learn new tasks: shop floor performance effects of knowledge transfer and performance feedback. J. Oper. Manag. **30**(3), 221–236 (2012). https://doi.org/10.1016/j.jom.2011.11.001
6. Grosse, E.H., Glock, C.H., Jaber, M.Y.: The effect of worker learning and forgetting on storage reassignment decisions in order picking systems. Comput. Ind. Eng. **66**(4), 653–662 (2013). https://doi.org/10.1016/j.cie.2013.09.013
7. Grosse, E.H., Glock, C.H.: The effect of worker learning on manual order picking processes. Int. J. Prod. Econ. **170**, 882–890 (2015). https://doi.org/10.1016/j.ijpe.2014.12.018
8. Chase, R.B., Aquilano, N.J., Jacobs, F.R.: Operations Management for Competitive Advantage. Mc-Graw Hill, Boston (2004)
9. U.S. Office of Personnel Management Federal Wage System Job Grading Standard for Sewing Machine Operating, 3111. TS-15 January 1971 Federal Wage System Job Grading System
10. Niebel, B.W., Freivalds, A., Niebel, B.W.: Methods, Standards, and Work Design, vol. 11. McGraw-Hill, Boston (2003)
11. Nembhard, D.A.: Heuristic approach for assigning workers to tasks based on individual learning rates. Int. J. Prod. Res. **39**(9), 1955–1968 (2001). https://doi.org/10.1080/00207540110036696
12. Uzumeri, M., Nembhard, D.: A population of learners: a new way to measure organizational learning. J. Oper. Manag. **16**(5), 515–528 (1998). https://doi.org/10.1016/S0272-6963(97)00017-X
13. Wright, T.P.: Factors affecting the cost of airplanes. J. Aeronaut. Sci. **3**(4), 122–128 (1936). https://doi.org/10.2514/8.155
14. Linstone, H.A., Turoff, M.: Delphi: a brief look backward and forward. Technol. Forecast. Soc. Chang. **78**(9), 1712–1719 (2011). https://doi.org/10.1016/j.techfore.2010.09.011

15. Yousuf, M.I.: Using experts' opinions through Delphi technique. Pract. Assess. Res. Eval. **12**(4), 1–8 (2007)
16. Nembhard, D.A.: The effects of task complexity and experience on learning and forgetting: a field study. Hum. Factors **42**(2), 272–286 (2000). https://doi.org/10.1518/001872000779656516
17. Nembhard, D.A., Osothsilp, N.: Task complexity effects on between-individual learning/forgetting variability. Int. J. Ind. Ergon. **29**(5), 297–306 (2002). https://doi.org/10.1016/S0169-8141(01)00070-1
18. Dar-El, E.M.: HUMAN LEARNING: From Learning Curves to Learning Organizations. Springer, New York (2000). https://doi.org/10.1007/978-1-4757-3113-2
19. Anzanello, M.J., Fogliatto, F.S.: Learning curve models and applications: literature review and research directions. Int. J. Ind. Ergon. **41**(5), 573–583 (2011). https://doi.org/10.1016/j.ergon.2011.05.001
20. Karger, D.W., Bayha, F.H.: Engineered Work Measurement. Industrial Press, New York (1966)
21. McCampbell, E.W., McQueen, C.W.: Cost estimating from the learning curve. Aero Digest **73**(4), 36–39 (1956)
22. Crossman, E.R.F.W.: A theory of the acquisition of speed-skill*. Ergonomics **2**(2), 153–166 (1959). https://doi.org/10.1080/00140135908930419
23. Tokunaga, M., Hashimoto, Y., Watanabe, S., Nakanishi, R., Yamanaga, H.: Methods for improving the predictive accuracy using multiple linear regression analysis to predict the improvement degree of functional independence measure for stroke patients. Int. J. Phys. Med. Rehabil. **5**(414), 2 (2017)
24. Oguntunde, P.G., Lischeid, G., Dietrich, O.: Relationship between rice yield and climate variables in southwest Nigeria using multiple linear regression and support vector machine analysis. Int. J. Biometeorol. **62**(3), 459–469 (2018)

A Study on Constructing KM System for Laboratories Based on the Three-Stage EDIS Spiral

Bingfei Tian ⓘ, Jianwen Xiang ⓘ, Ming Yang ⓘ, Dongdong Zhao ⓘ, and Jing Tian ✉ ⓘ

School of Computer Science and Technology, Wuhan University of Technology, Wuhan, China

{tbf,myang,zdd,jtian}@whut.edu.cn, xiangjw@gmail.com

Abstract. With development of information technology, knowledge management has become an indispensable matter in the field of academia. Colleges and universities have plenty of opportunities to use knowledge management to support their tasks. Knowledge management not only manages the knowledge and information technology inside and outside the organization, but also needs to carry out knowledge management on the organization, such as on organizational structure, human resources, organizational culture and other aspects, in order to achieve the established knowledge management objectives. Knowledge management is multi-faceted. A knowledge management system is required to integrate multi-faceted knowledge management. Based on the questionnaire collected, students' implicit barriers and special requirements in their self-learning were analyzed and found. Specifically according to the three-stage EDIS spiral model proposed by Sun et al. in 2018, there are three stages, which are development of research plan, implementation of research and dissemination of research results, and then we designed the function module for every stage. Combined the relationships among members of the organization, including teacher-student relationship and classmate relationship, with the users' specific needs, we designed this scientific and reasonable KM system which is suitable for science and engineering laboratory. The results show that our system can help laboratory members to use knowledge resources efficiently and improve the level of knowledge management of personal and laboratory in a certain extent.

Keywords: Knowledge management · The three-stage EDIS spiral Laboratory management system

1 Introduction

With the outbreak of information, the information construction of universities and colleges has been put on the agenda, and it has become a top priority for all scientific research institutions [1]. At the same time, with the development of research work in major universities and colleges, knowledge management has been particularly important [2]. Having a reasonable knowledge management system will greatly increase the efficiency of scientific research and reduce unnecessary waste of time [3].

© Springer Nature Singapore Pte Ltd. 2018
J. Chen et al. (Eds.): KSS 2018, CCIS 949, pp. 237–251, 2018.
https://doi.org/10.1007/978-981-13-3149-7_18

The theory of knowledge management was first proposed in the early 1990s. Then, academics and entrepreneurs turn their research direction into knowledge management [4]. This paper defines knowledge management in academia as any systematic activity related to supporting and enhancing the creation of scientific knowledge and realizing the research target, including social processes and related computer technology tools [5].

In the era of knowledge economy, knowledge management has always been one of the most important strategic actions in an organization [6]. Knowledge is at the center stage of knowledge management practice and involves human participation [7]. Some seniors pointed out: "The most valuable productivity in the new era should be talented people with relevant knowledge reserves."

Research institutions and universities are important components of the social academic community. One of the most important roles is to create and disseminate scientific knowledge. This is the fundamental driving force for social progress and economic development [8]. Therefore, the enhancement of the scientific knowledge creation process in academia will largely influence the emergence of the knowledge economy [9]. Research labs are the cornerstone of current research. Knowledge management is a very advanced management model. It has high value for use in protecting the research results of laboratories, realizing knowledge creation and knowledge sharing, and establishing mutual research laboratories [10]. Therefore, we introduce KM to laboratory, to improve the efficiency of innovation and also to enhance the laboratory's core competitiveness.

This paper starts with the relevant theories and models, such as the three-stage EDIS spiral, which the study base on, and through introducing this model, we explain the motivation of building a knowledge management system. Specifically speaking, we analyzed the three-stage EDIS spiral and researched the users' requirements combining with the questionnaire. After summarizing the problem exiting in the graduate students, we designed the corresponding function module, and integrate the KM system, to help members of the laboratory make rational use of knowledge resources and further improve the efficiency of knowledge innovation. What's more, we payed a return visit to the users, and found the preponderance and deficiency of the system. Through the feedback we proved the practicability and validity of the system.

2   Relevant Theories and Models

2.1   SECI Model (Organizational Knowledge Creation)

Nonaka and Takeuchi put forward a new knowledge creation model called SECI model in 1995 [11]. This model is one of the most well-known theories in organizational knowledge creation [12]. It defines organizational knowledge creation as the spiral transformation process of individual tacit knowledge to organizational dominant knowledge. It consists of three components: SECI, Ba, and Knowledge Assets, which are interacted organically and dynamically. The knowledge assets of the organization are shared among Ba by the members of the organization. Meanwhile, the individual tacit knowledge of the members is transferred and amplified through SECI among Ba [3].

SECI means four creative processes of knowledge: 1. Transform the individual tacit knowledge into the organization tacit knowledge (Socialization); 2. Transform the organization tacit knowledge into the organization explicit knowledge (Externalization); 3. Transform the organization explicit knowledge into the individual explicit knowledge (Combination); 4. Transform the individual explicit knowledge into individual tacit knowledge (Internalization). Ba, the meaning of "field" in Japanese, refers to the place where knowledge is converted, including certain times and places. The fields corresponding to the above four processes are Origination Ba, Dialoguing Ba, Systemizing Ba and Exercising Ba (see in Fig. 1).

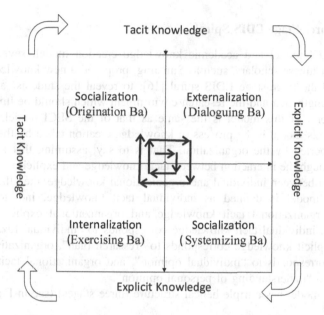

Fig. 1. The SECI model (Nonaka and Takeuchi, 1995)

2.2 Academic Knowledge Creation

In 2006, Wierzbicki and Nakamori put forward the academic knowledge creation to explain the knowledge creation processes of graduate students in different fields [13]. There are three modes of normal academic knowledge creation: Hermeneutics, which is expressed as EAIR (enlightenment-analysis-immersion- reflex) spiral; Debate, expressed as EDIS (enlightenment-debate-immersion-select) spiral; Experiment, expressed as EEIS (enlightenment-experiment-interpretation-selection) spiral (see in Fig. 2).

The organizational theory that uses knowledge emphasizes the role of tacit knowledge for specific situations and specific organizations [14]. Achieving a constant shift from one type of knowledge to another is the basis for creating new knowledge resources in an organization [15].

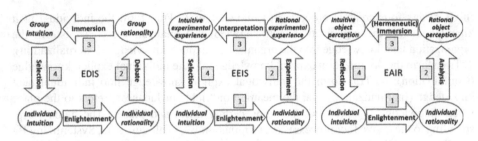

Fig. 2. Three academic knowledge creation models (Wierzbicki and Nakamori, 2006)

2.3 The Three-Stage EDIS Spiral

Based on SECI model and academic knowledge creation model, several Japanese scholars and Chinese scholars, such as Sun Jing, proposed a new knowledge creation model, called the three-stage EDIS spiral [16], to reveal the students' ability which should be strengthened and the research environment which should be improved. The knowledge form of this model is the same as that of the SECI model, but what it expresses in this model is the process of knowledge creation taken by the individuals who are supported by the organization. That is to say, assuming that knowledge is generated through the interaction between tacit knowledge and explicit knowledge and the interaction between individual and organizational knowledge, then the knowledge form in the model is defined as individual tacit knowledge, individual explicit knowledge, organizational tacit knowledge and organizational explicit knowledge. Among them, individual tacit knowledge corresponds to "individual fuzzy concept", individual explicit knowledge corresponds to "research plan", organizational explicit knowledge corresponds to "individual opinion", and organizational tacit knowledge corresponds to "understanding of personal opinion".

This new model has a triple helical structure (three stages) (see in Fig. 3).

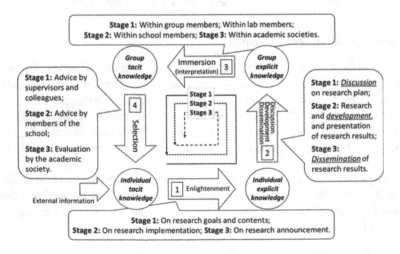

Fig. 3. The three-stage EDIS spiral (Sun et al., 2017)

The first lab of the spiral expresses the process of planning study. Through reading a large number of literature and discussing with mentor and predecessor, the students determine the research goal and the concrete research content confidently. In the process, the most important thing is "discussion."

The second lab of the spiral expresses the process of implementing the study. Through the experiment, computer simulation and questionnaire survey, they carry out the research in detail and then obtain research results.

The third lab of the spiral expresses the process of spreading the research results. Through submitting the journal articles and publishing papers in the academic conference, they make the research results understood and accepted by academics.

So a steady stream of new ideas and knowledge are created and verified in the spiral. Knowledge creation process (1 and 2) and the validation process (3 and 4) are included in this model.

3 Summary and Findings

This study uses the survey work of Sun and Tian in 2017, in which they found some questions existing in the ability and environment [17]. We analyze and summarize the previous survey results, and it can help us analyze the potential and actual needs of graduate students and teachers for knowledge management systems, and find out from which aspect the knowledge management systems can improve research efficiency. Finally we come up with three findings.

3.1 Problem in Stage 1

Higher education can not let students learn critical thinking, quantitative reasoning and research ability, so, some graduate students need to have appropriate research process model for them. The first stage of the three-stage EDIS spiral explains the process of creating a plan. Before deciding the research direction, the members need to read a lot of references and discuss with their teachers to get more confidence and get the plan. Also, they need to know if there are some academic conferences suitable for his direction and to judge if the time is enough. Analyzing from the survey results, we can find that the students don't think highly of the present study environment and their ability to begin a study by themselves. As for knowledge access, the way is very scattered and the efficiency is very low. The students say that they need enough raw data and detailed guidance, and they need their teachers provide necessary references.

3.2 Problem in Stage 2

After accepting teachers' suggest, the students should write their detailed plan to start the actual study, and in the process of study, they should submit report to show their progress and results, in order to get more advices from the teacher to improve the study. Also, the students should learn to understand and choose some useful advices. This process is the process of transforming organization tacit knowledge into individual tacit knowledge. In the survey result, we can find that the students' confidence to research is low, the self-assessment of research capacity is low, and they couldn't get timely guidance.

3.3 Problem in Stage 3

After finishing the study, the student should write an academic paper and publish to relevant journals or conferences, to spread his research result to public. On the conference, he must answer the questions and advices put forward by the reviewers. And it has very important meaning for his following research. However, student needs a great deal of self-confidence and grasp to complete the report, which are precisely what today's graduate student lacked.

The results of the survey help us to analyze the potential needs or actual needs of graduate students and their teachers for KM system, and to find how to promote efficiency of studying.

4 Function Module Design and System Construction

In Sect. 2, we introduced the Three-stage EDIS Spiral, it's this study's theoretical foundation. In Sect. 3, we introduced the existing problems of graduate students in scientific research, it's the study's realistic demand. In this section, we combine the theoretical foundation with the practical needs, and design corresponding modules to solve the problems by combining the relevant knowledge of computer science.

4.1 Function Module Design

The Stage of Creating a Research Plan. To solve the problems mentioned, the module for recommending journals and conferences and the module for publishing papers are proposed. This module helps researchers obtain the background and expertise needed to write a new research project application more easily. This module is mainly related to the tutor or research leader. In almost all research activities, except the research topic which is mainly proposed by the tutor, the knowledge source is mainly obtained through self-study, then the tutor classmate, and finally seeking external help. The help from the mentor is mainly in the writing of the article. The former also reveals the structure of a stuff, that is, the person in charge and the graduate student. Here, the members and teachers sort out the journals and conferences which are useful for studying of their laboratory, and then recommend to other members. Recommended content includes the theme of the conferences and the date of the deadline, in order to be convenient for students to choose the next step of research contents, and have a preliminary schedule. At the same time, they can upload their paper published before to paper database of the system. What's more, the teacher could put forward his views and research guidance, and point out what is the study's instruction meaning for the development of laboratory and individual. As the effective way for knowledge assets of members, these provide the early knowledge reserve for students and prompt them to create a clear study plan.

The Stage of Implementing the Study. Taking into account above problems in this stage, the module for plan and summary is proposed. Here, students record their own study plan of next phase, sum up their work completed at this stage and experience, and other members and teachers of the laboratory can comment on the plan and summary. Through recording the plan and summary and viewing other members' plan and summary, the students could understand themselves and the current progress of the study in the group better, at the same time, have a clear goal about the work content of next phase. By analyzing other members' and teachers' comments and guidance, they modify their own plans and summary, and then discover their mistakes and short-comings or their advantages, thereby improve their research ability and research confidence. What's more, in this module, the student could set the schedule reminder to achieve his plan in time.

The Stage of Spreading the Research Results. Based on the problems in this stage, we design the module for publishing papers and the module for seminar. Here, after finishing the paper, the students upload their papers to provide to other members and teacher to read. At the seminar, everyone could put forward some questions about the study and paper, and they must answer carefully one by one. It is also like a small simulation report. They can also initiate a video conference. By this way, students could be aware of their shortcomings clearly, make up for it and have a better performance on the conference.

Other Functions. Except these, an integral system also needs other functions to make itself more practical. So, we design the timely communication function to make it more convenient for communication, the module of lab members profile, search function, and so on, (see in Fig. 4).

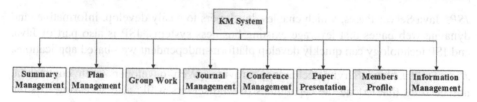

Fig. 4. The system function module diagram

4.2   System Construction

This project takes the Trusted Computing Laboratory as a research example, which is in the School of Computer Science and Technology of Wuhan University of Technology. According to the survey results and discussions with the laboratory members, we analyzed their specific needs, and then designed and implemented the laboratory knowledge management system for them to improve the knowledge management level of the laboratory and the process of scientific research and innovation effectively.

Introduction to Related Technologies and Tools. This topic is for the design and implementation of the laboratory knowledge management system, and finally realizes an interface introduction and a fully functional website system. The system is a web based system. Basically, users only need a web browser, such as Internet Explorer, and an Internet link to access it. It uses the SSH framework as a whole architecture, which makes the system easier to expand and maintain, so that the system can be continuously updated and improved according to the changes of the needs of the laboratory members, thereby improving the practicability of the system. The following are the main techniques and tools used to develop the system:

SSH Framework. The SSH framework is actually a collection of three frameworks, including struts, spring, and hibernate frameworks. It is an open source integration framework commonly used in current development to build flexible, easy-to-expand multi-layer web applications. The system is divided into four layers: responsibilities layer, business logic layer, data persistence layer and domain module layer (physical layer).

Bootstrap. This is a development framework commonly used by front-end developers. It is based on HTML, CSS, and JAVASCRIPT. It is simple and convenient to use, and shortens the development cycle in a certain sense.

Java and Java Applet. Java was developed by Sun Company in 1995. Applet is a small program defined by Java. It is positioned to perform certain special tasks, such as animated graphics and interactive tools, in other windows such as browsers. For running systems, client users need to download some Java plugins.

Tomcat. This system uses Tomcat 5.1 as the web server. Tomcat is a lightweight server developed by Sun Company.

Struts. Struts is an auxiliary development tool for design patterns. It provides the MVC pattern on the development website. It also provides a set of support classes.

JSP. Java Server Pages, which enables developers to easily develop, informative, and dynamic web pages that leverage existing business systems. JSP is also part of Java, and JSP technology can quickly develop platform-independent web-based applications.

Java Servlet. Java Servlet technology provides Web developers with convenient, unified rules that extend the capabilities of web servers while also accessing existing business logic.

XAMPP. Apache + MySQL + PHP + PERL, which is a very powerful integrated software tool.

Feasibility Analysis. The feasibility analysis is a simple evaluation of the project before development [18]. This paper analyzed the system from the following three aspects:

Technical Feasibility. This system is a system for database knowledge management and query. Mainly using JSP technology, SSH framework and MySQL database, these technologies are very mature now, and the laboratory also provides a complete development specification. Therefore, it is entirely feasible to develop technical feasibility.

Economic Feasibility. Most colleges and graduate students already have certain hardware requirements for computer equipment. This software is developed on my own computer system and does not require additional hardware cost. The software is all open source free development system, and no cost investment. Overall, the project team does not need to invest a lot of money, and the system can create great value after development. In summary, the system is completely economically viable.

Operational Feasibility. This system is designed and implemented with JSP technology. Users only need to open the browser input URL to enter the system. The account is registered by the method prompted by the system, and then log in to enter the home page. Click on the function you need to use the system. The interface of the system is simple and easy to operate. The system can meet the operational requirements of the lab members. The design of a SSH-based laboratory knowledge management system is operationally feasible.

Log in to the system, you can open the home page of the lab management website (see in Fig. 5).

Fig. 5. The home page of the lab management system

5   Evaluation

5.1   User Feedback

After completed, the system is handed over to the lab members for preliminary use, and we design a few simple questions as user feedback, so as to help ourselves know the system's shortcomings and further improve it. Here is the composition of the laboratory (see in Table 1).

Table 1. Composition of the laboratory

Total number of members: 28			
Teacher	4	Teaching experience ($\geq$ 3 years)	3
		Teaching experience (>3 years)	1
PhD students	4	Paper writing experience	4
Masters	18	Paper writing experience	10
		Writing now	4
		Never writing	4

For student users:

Q1: What is your overall impression of the lab knowledge management system, such as design style and layout.

Q2: Do you log in and view this website frequently? How many times do you visit this website in a week?

Q3: Which website module you use most often, and give your reasons.

Q4: What do you think is the most obvious change you have brought to the system?

Q5: From August 2016 to July 2017, how many papers did you finish? And after using our system, that is to say, from August 2017 to July 2018, how many paper did you write? What's more, how long does it take you to finish a paper before and after using the system?

Q6: Please give some suggestions to our system from the perspective of student users.

For teacher users:

Q7: Do you log in and view this website frequently? How many times do you visit this website in a week?

Q8: What convenience do you think brings to the system?

Q9: What changes have you made to your students after using the system?

Q10: Please give some suggestions to our system from the teacher's point of view.

Based on the feedback given by the user, we made the following summary:

1. The user's overall impression of the website is good, and eight users all said that the page is simple and practical, and the function is easy to operate, which can be seen at a glance. But at the same time, five users have raised the problem that the interface is too rigid.

2. The user logs into the system mainly to record his own research plan and summarize his work, and based on the plans and conclusions of other members, to understand the progress of the project, analyze and improve his plan; Share your current problems and seek help from lab members;

3. The student users said that the most useful part is the discussion area, which provides an online communication space for mutual exchange and exchange, and avoids the shyness and embarrassment that may occur when face-to-face communication, which brings convenience to internal communication. And some users

indicated that they prefer to plan and summarize modules. At the same time, all users feel that the small function of uploading files is very practical.

4. The student users proposed in the feedback that the knowledge management system enhances the standardization of knowledge, provides a standard document template for more important documents, and improves the convenience and standardization of work.

5. According to the teachers' feedback, after using this system, they can always see the progress of the students' research and supervise them. They can clearly find out the problems students have in the research, so as to contact the students and help them to modify them. The students have become more active and will respond positively to the comments given by the teachers and tell their reasons and opinions. Since the laboratory began using the system in August 2017, the number of papers published between August 2016 and July 2017 are compared here with the number of papers published by the laboratory members during the period from August 2017 to July 2018, one year after the use of the system. At the same time, we sorted out the time required for PhD students and masters to complete a paper before and after using the system, from the time the idea was presented to the time the paper was completed. We found that, after using KM system, the number of paper published was increased a little, simultaneously the time required to complete a paper was greatly reduced.

6. The users put forward constructive opinions on the functions of the system. For example, the function of adding files in the group discussion area can make the sharing more convenient, and the schedule reminding function is enabled to prompt the users to complete their plans in time.

5.2 Comparison

Sun and Tian have designed a question list to evaluate research ability and research environment in 2017 [17]. The researchers introduced four questions to assess the capabilities and environment of each process: ability of doing research, importance of ability, environment of doing research and the necessity of the environment. And after students using our system, we asked them to re-fill part of this questionnaire. The meanings of the levels from 1 to 5 are, for example (see in Fig. 6), as follows: 1 = strongly dissatisfied; 2 = dissatisfied; 3 = neither; 4 = satisfied; and 5 = strongly satisfied. The comparison of results helps us to verify the effectiveness of the system.

By analyzing the survey data and comparing it with their previous results, we found that the students' ability of doing research and importance have been enhanced, especially for enlightenment in Stage 1, development in Stage 2 and selection in Stage 3 (see in Fig. 7).

We also found that the environment of doing research and necessity have been improved, especially for discussion in Stage 1, discussion in Stage 2 and immersion in Stage 3 (see in Fig. 8). The comparison of results helps us to verify the effectiveness of the system.

Stage 1: **Ability** of Enlightenment (1-E-A)	
Questions for Students	**Questions for Teachers**
A) Are you satisfied with your ability to plan your research? (Do you have a lot of ideas and experience to make your research plan?) B) Do you think this ability is important for you?	A) Are you satisfied with your students' ability to plan their research? (Do your students have a lot of ideas and experience to make research plans?) B) Do you think this ability is important for students?

A)	Dissatisfied	1	2	3	4	5	Satisfied	A)	Dissatisfied	1	2	3	4	5	Satisfied
B)	Unimportant	1	2	3	4	5	Important	B)	Unimportant	1	2	3	4	5	Important

- If your answer is 'dissatisfied' but 'important', what would you like to do specifically?
- If your answer is 'unimportant', please write the reason.

- If your answer is 'dissatisfied' but 'important', what would you like to ask your students specifically?
- If your answer is 'unimportant', please write the reason.

Fig. 6. An example of evaluation sheet (Sun et al., 2017)

Fig. 7. Comparison of ability and importance

5.3 Limitations and Future Work

Through the summary of user feedback, this paper found that the system has certain defects and shortcomings in design and function.

(1) The interface design is too simple, although practical, but not beautiful enough or even dull, not very attractive to users.

(2) The functions realized by the system are relatively simple, only meet the basic needs of users, and many personalized and user-friendly functions are not implemented, such as adding a memo function to remind users to complete tasks on time,

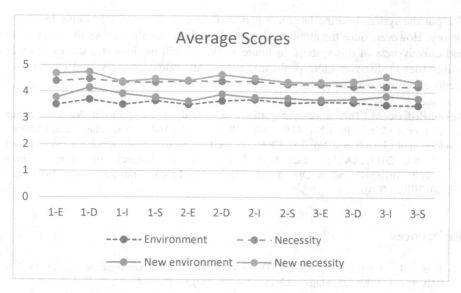

Fig. 8. Comparison of environment and necessity

designing instant messaging functions for users. The connection is more convenient and so on.

(3) The current design of the system can only meet the needs of specific users. If it is necessary to expand the scope of application, a wide range of investigation and analysis is required to design a knowledge management system suitable for most laboratories.

In the future, we will deepen understanding and learn about knowledge management, combine it with computer expertise to form a systematic theory and ideas, and re-examine the system of this design to continuously improve system functions, with a view to giving laboratory knowledge. Management brings greater benefits. After the system is perfected, it will be put into other laboratories for a period of time, and then a systematic investigation and analysis will be carried out to prove the practicability of the system with the final data.

6 Conclusion

This paper first introduced the development of knowledge management in the era of knowledge economy. And then we explained the motivation of building a knowledge management system through the relevant theories and models. After this, we analyzed the survey results proposed by Sun [14], summarized the problems exiting in graduate students, analyzed their needs for knowledge management system, and then designed the system modules. After determining the function modules, we realize the designing and implementation of the KM system in the WEB server by the relative technologies of software engineering, such as Java, SSH framework, MySQL and Photoshop.

We put the system into the lab for a period of time, and conduct a simple interview survey. However, since the number of respondents is too small to prove the practicality and effectiveness of this system. In future work, we will implement a survey in other universities to discover more problems, expand the range of users, and collect user feedback to improve the system.

Acknowledgement. This work was partially supported by the National Natural Science Foundation of China (Grant No. 61672398, 61806151), the Hubei Provincial Natural Science Foundation of China (Grant No. 2017CFA012), the Key Technical Innovation Project of Hubei (Grant No. 2017AAA122), the Applied Fundamental Research of Wuhan (Grant No. 20160101010004), and the Open Fund of Hubei Key Lab. of Transportation of IoT (Grant No. 2017III028-004).

References

1. Roger, F., Jennifer, R., Rachel, D.: Knowledge sharing amongst academics in UK universities. J. Knowl. Manag. **17**(1), 123–136 (2013)
2. Wierzbicki, A.P., Nakamori, Y.: Creative Environments: Issues of Creativity Support for the Knowledge Civilization Age. Springer, Heidelberg (2006). https://doi.org/10.1007/978-3-540-71562-7
3. Ren, H., Tian, J., Nakamori, Y., Wierzbicki, A.P.: Electronic support for knowledge creation in a research institute. J. Syst. Sci. Syst. Eng. **16**(2), 235–253 (2007)
4. Zack, M.H.: Developing a knowledge strategy. Calif. Manag. Rev. **41**(3), 125–145 (2002)
5. Wu, J., Tian, J., Shi, B., Xiang, J.: A study of knowledge management and knowledge creation in undergraduate based on surveys in a Chinese research university. In: The 16th Annual International Symposium on Knowledge and Systems Sciences, pp. 60–68. Springer, Xi'an (2015)
6. Danate, M.J., Canales, J.I.: A new approach to the concept of knowledge strategy. J. Knowl. Manag. **16**(1), 22–44 (2012)
7. Sun, J., Nakamori, Y., Tian, J., Xiang, J.W.: Exploring academic knowledge creation models for graduate researches. In: IEEE International Conference on Software Quality, Reliability and Security Companion, pp. 202–209. IEEE, Vienna (2016)
8. Roger, F., Jennifer, R., Rachel, D.: Knowledge sharing amongst academics in UK universities. J. Knowl. Manag. **17**(1), 123–136 (2013)
9. Wu, W.L., Lee, Y.C., Shu, H.S.: Knowledge management in education organization: a perspective of knowledge spiral. Int. J. Organ. Innov. **5**(4), 7–13 (2013)
10. Wu, J., Tian, J., Lu, L., Weng, C., Xiang, J.: Surveys on knowledge management and knowledge creation in a Chinese research university. In: The 2016 IEEE International Conference on Software Quality, Reliability and Security Companion, pp. 186–193. IEEE, Vienna (2016)
11. Nonaka, I., Takeuchi, H.: The Knowledge-Creating Company: How Japanese Companies Create the Dynamics of Innovation. Oxford University Press, Oxford (1995)
12. Wierzbicki, A.P., Nakamori, Y.: Creative environments: issues of creativity support for the knowledge civilization age. J. Doc. **65**(3), 523–527 (2007)
13. Wierzbicki, A.P., Nakamori, Y.: Creative Space: Models of Creative Processes for the Knowledge Civilization Age. Springer, Heidelberg (2006). https://doi.org/10.1007/b137889
14. Li, B., Zhang, J., Zhang, X.: Knowledge management and organizational culture: an exploratory study. Creative Knowl. Soc. **3**(1), 56–69 (2013)

15. Tian, J., Nakamori, Y., Wierzbicki, A.P.: Knowledge management and knowledge creation in academia: a study based on surveys in a Japanese research university. J. Knowl. Manag. **13**(2), 76–92 (2009)
16. Sun, J., Tian, J., Huynh, V.N., Nakamori, Y.: Modeling the knowledge creation process of graduate research. J. Syst. Sci. Syst. Eng. (2018)
17. Jing, S., Jing, T., Van-Nam, H.: Knowledge management in graduate research. In: 2017 IEEE International Conference on Software Quality, Reliability and Security Companion (QRS-C), pp. 480–485. IEEE, Prague (2017)
18. Sarnikar, S., Deokar, A.V.: A design approach for process-based knowledge management systems. J. Knowl. Manag. **21**(4), 693–717 (2017)

Some q-Rung Orthopair Fuzzy Dual Maclaurin Symmetric Mean Operators with Their Application to Multiple Criteria Decision Making

Jun Wang[1], Runtong Zhang[1(✉)], Li Li[1], Xiaopu Shang[1], Weizi Li[2], and Yuan Xu[1]

[1] School of Economics and Management, Beijing Jiaotong University, Beijing 100044, China
{14113149, rtzhang, 16120608, sxp, 17120627}@bjtu.edu.cn
[2] Informatics Research Centre, Henley Business School, University of Reading, Reading RG6 6UD, UK
weizi.li@henley.ac.uk

Abstract. This paper investigates multiple criteria decision making (MCDM) with q-rung orthopair fuzzy information. Recently, some aggregation operators have been developed for q-rung orthopair fuzzy sets (q-ROFSs). However, the main flaw of these operators is that they fail to capture the interrelationship among multiple input arguments. The dual Maclaurin symmetric mean (DMSM) is an efficient aggregation function which can reflect the interrelationship among multiple input variables. Motivated by the dual Maclaurin symmetric mean (DMSM), we extend DMSM to q-ROFSs and propose some q-rung orthopair fuzzy dual Maclaurin symmetric mean operators. We also investigate the properties and special cases of these operators. Further, a novel approach to multiple criteria decision making (MCDM) is introduced. We apply the proposed method in a best paper selection problem to demonstrate its effectiveness and advantages.

Keywords: q-rung orthopair fuzzy set · Dual maclaurin symmetric mean
q-rung orthopair fuzzy dual maclaurin symmetric mean
Multiple criteria decision making

1 Introduction

With the rapid development of economy, society and technologies, complexity of decision making problems has been gradually increasing during the past several decades. In day-to-day decision making, we have to consider not only external factors, but also the subjective will of decision makers. MCDM has been proved to be a very effective decision making model and owing to its efficiency, it has been successfully applied in quite a few fields, such as investment selection, airline evaluation, supplier selection and medical diagnosis. MCDM has strong ability of modelling the process of practical decision making problems and so that it has drawn much scholars' attention.

© Springer Nature Singapore Pte Ltd. 2018
J. Chen et al. (Eds.): KSS 2018, CCIS 949, pp. 252–266, 2018.
https://doi.org/10.1007/978-981-13-3149-7_19

Basically, MCDM is an activity to select the best alternative from a set of candidates under decision makers' assessments on a group of criteria. Due to the increased complexity in social economics and management, we always face the difficulties of representing criterion values in complicated and fuzzy decision making environments. Thus, more and more scholars paid their attention to tools that can handle fuzziness and uncertainty effectively. Up to now, quite a few theories that can deal with fuzzy information have been proposed. For example, Atanassov [1] proposed the concept of intuitionistic fuzzy set (IFS) for describing vagueness and ambiguity of information. In contrast to fuzzy sets [2], IFSs have a stronger advantage in describing the inherent uncertainty of information itself. Since IFSs can better describe human cognitive processes, they have drawn much attention and has been successfully applied to MCDM [3–6]. However, the constraint of IFSs cannot be satisfied in many complicated decision making problems. In other words, the qualification conditions of IFS limit its scope of use. For instance, a decision maker may provide 0.6 for membership degree and 0.7 for non-membership degree respectively. We notice $0.6 + 0.7 = 1.3 > 1$, so that the decision value cannot be represented by IFSs. To effectively address this situation, a concept of Pythagorean fuzzy set (PFS) has been introduced by Yager [7]. Evidently, PFS relaxes the constraint of IFS. That is to say, the constraint of PFSs is broader than that of IFSs. This characteristic makes PFS more powerful and it can describe more information range and capture more information in the process of MCDM.

After decision makers evaluate all the criterion values in MCDM, the most urgent issue is how to make an option based on expert's assessments. Generally speaking, the methods to MCDM includes two types, one is based on traditional decision making methods, and another one is based on aggregation operators. For instance, Zhang and Xu [8], Liang et al. [9], and Perez-Dominguez [10] respectively extended the traditional TOPSIS and MOORA to MCDM within a Pythagorean fuzzy context. Moreover, quite a few Pythagorean fuzzy aggregation operators have been presented. Xing et al. [11] developed some series of Pythagorean fuzzy Choquet integral operators based on Frank t-conorm and t-norm. Zhang et al. [12] extended the Pythagorean fuzzy Bonferroni mean operators to make them capable to capture the interrelationship among all Pythagorean fuzzy numbers (PFNs). Further, Xu et al. [13] proposed some novel operators which not only consider the interaction among membership and non-membership degrees, but also reflect the interrelationship among aggregated PFNs.

As time goes on, decision making problems become more complex and diverse, and many special cases arise. Although the IFSs and the PFSs can solve MCDM effectively, there are still problems that cannot be solved by the two theories. For example, a decision maker provides (0.8, 0.9) as the criteria value of a certain alternative. As $0.8 + 0.9 = 1.7 > 1$, and $0.8^2 + 0.9^2 = 1.45 > 1$, then the criteria value (0.8, 0.9) is note valid for IFSs and PFSs, whereas can be represented by q-ROFSs, which are a good tool proposed by Yager [14]. The q-ROFSs are efficient to address situations in which the sum of qth power of the membership degree and the qth power of the degree of the membership is equal to or less than one. In contrast to IFSs and PFSs, it can be seen that the q-ROFSs have a wider range of applications, can describe more complex fuzzy information, and bring less information distortion. Recently, the problem of q-rung orthopair fuzzy MCDM has increasingly become a research hotspot in academia, and has achieved some important research results, among which the operators proposed by Liu and Wang [15], Liu and Liu [16], Liu and Wang [17], and Wei et al. [18] are the most fundamental and significant.

The aforementioned q-rung orthopair fuzzy aggregation operators can be successfully applied in MCDM. However, the main shortcoming of the proposed operators is that they overlook the interrelationship between input arguments (i.e. [15]) or can only capture the interrelationship between any two aggregated variables (i.e. [16–18]). So these operators and methods can't make the right decisions. In most practical decision making problems, interrelationship exists among multiple criteria. Thus, when calculating the overall assessment values of alternatives, the interrelationship among multiple arguments should be considered. This is the main motivation of this work. The recently proposed Maclaurin symmetric mean [19] is an effective aggregation operator that can capture the interrelationship among multiple aggregated arguments. Owing to its advantages and superiorities for fusing information, MSM has been effectually applied in IFSs, PFSs, and hesitant fuzzy sets. More recently, Qin and Liu [20] proposed a dual form of MSM, called DMSM and applied it in fusing hesitant fuzzy information. Motivated by DMSM, this paper extended DMSM to q-ROFSs and proposed some novel q-rung orthopair fuzzy aggregation operators. In addition, we utilize the proposed operators to solve MCDM in a q-rung orthopair fuzzy environment.

The main motivations and objectives of this paper are: (1) to propose some new q-rung orthopair fuzzy aggregation operators which reflect the interrelationship among multiple aggregated arguments, (2) to introduce a novel approach to MCDM. The remainder of the paper is organized as follows. Section 2 briefly recalls some basic concepts. Section 3 presents some q-rung orthopair fuzzy dual Maclaurin symmetric mean operators. Section 4 introduces a novel approach to MCDM with q-rung orthopair fuzzy information. Section 5 provides a numerical example to illustrate the effectiveness of the proposed method. Section 6 summarizes the paper.

2 Basic Concepts

In this section, we review some basic concepts which will be used in the following sections.

2.1 The q-Rung Orthopair Fuzzy Set

Definition 1 [14]. Let X be an ordinary set, then a q-ROFS A is defined as follows:

$$A = \{\langle x, u_A(x), v_A(x)\rangle \,|x \in X\}, \tag{1}$$

where $u_A(x)$ and $v_A(x)$ denote the membership and non-membership degrees respectively, satisfying $0 \le u_A(x) \le 1$, $0 \le v_A(x) \le 1$ and $0 \le u_A(x)^q + v_A(x)^q \le 1, (q \ge 1)$. Liu and Wang [15] called $(u_A(x), v_A(x))$ a q-rung orthopair fuzzy number (q-ROFN), which can be denoted as $\alpha = (u, v)$.

Definition 2 [15]. Let $\alpha_1 = (u_1, v_1)$ and $\alpha_2 = (u_2, v_2)$ be two q-ROFNs and λ be a positive real number, then

(1) $\alpha_1 \oplus \alpha_2 = \left(\left(u_1^q + u_2^q - u_1^q u_2^q \right)^{1/q}, v_1 v_2 \right)$,

(2) $\alpha_1 \otimes \alpha_2 = \left(u_1 u_2, \left(v_1^q + v_2^q - v_1^q v_2^q \right)^{1/q} \right)$,

(3) $\lambda \alpha_1 = \left(\left(1 - \left(1 - u_1^q \right)^\lambda \right)^{1/q}, v_1^\lambda \right)$,

(4) $\alpha_1^\lambda = \left(u_1^\lambda, \left(1 - \left(1 - v_1^q \right)^\lambda \right)^{1/q} \right)$.

Definition 3 [15]. Let $\alpha = (u, v)$ be a q-ROFN, then the score function of α is defined as $S(\alpha) = u^q - v^q$, the accuracy function of α is defined as $H(\alpha) = u^q + v^q$. Let $\alpha_1 = (u_1, v_1)$ and $\alpha_2 = (u_2, v_2)$ be any two q-ROFNs, $S(\alpha_1)$ and $S(\alpha_2)$ be the score functions of α_1 and α_2 respectively, $H(\alpha_1)$ and $H(\alpha_2)$ be the accuracy functions of α_1 and α_2 respectively, then

(1) if $S(\alpha_1) > S(\alpha_2)$, then $\alpha_1 > \alpha_2$;
(2) if $S(\alpha_1) > S(\alpha_2)$, then

> if $H(\alpha_1) > H(\alpha_2)$, then $\alpha_1 > \alpha_2$;
> if $H(\alpha_1) = H(\alpha_2)$, then $\alpha_1 = \alpha_2$.

2.2 Dual Maclaurin Symmetric Mean

The MSM is an aggregation technology proposed by Maclaurin [19] for crisp numbers. Recently, Qin and Liu [20] proposed a dual form of MSM, called the DMSM, which is defined as followings.

Definition 4 [20]. Let $a_j (j = 1, 2, \ldots, n)$ be a collection of crisp numbers, and $k = 1, 2, \ldots, n$, if

$$DMSM^{(k)}(a_1, a_2, \ldots, a_n) = \frac{1}{k} \left(\prod_{1 \le i_1 < \ldots < i_k < n} \left(\sum_{j=1}^{k} a_{i_j} \right)^{1/c_n^k} \right), \qquad (2)$$

then $DMSM^{(k)}$ is called the DMSM operator, where $(i_1, i_2, \ldots, i_k)$ traversal all the k-tuple combination of $(1, 2, \ldots, n)$, C_n^k is the binomial coefficient.

3 Some q-Rung Orthopair Fuzzy Aggregation Operators Based on DMSM

In this section, we extend DMSM to q-rung orthopair fuzzy environment and propose some novel aggregation operators.

3.1   The q-Rung Orthopair Fuzzy Dual Maclaurin Symmetric Mean (q-ROFDMSM) Operator

Definition 5. Let $\alpha_i = (u_i, v_i)(i = 1, 2, \ldots, n)$ be a collection of q-ROFNs, and $k = 1, 2, \ldots, n$. If

$$q - ROFDMSM^{(k)}(\alpha_1, \alpha_2, \ldots, \alpha_n) = \frac{1}{k}\left(\underset{1 \le i_1 < \ldots < i_k \le n}{\otimes} \left(\overset{k}{\underset{j=1}{\oplus}} \alpha_{i_j} \right)^{1/c_n^k} \right), \tag{3}$$

then $q - ROFDMSM^{(k)}$ is called q-ROFDMSM, where $(i_1, i_2, \ldots, i_k)$ traversal all the k-tuple combination of $(1, 2, \ldots, n)$, C_n^k is the binomial coefficient.

According to the operational rules of q-ROFNs presented in Definition 2, the following theorem can be derived.

Theorem 1. Let $\alpha_i = (u_i, v_i)(i = 1, 2, \ldots, n)$ be a collection of q-ROFNs, and $k = 1, 2, \ldots, n$. The aggregated value by using q-ROFMSM is still a q-ROFN and

$$q - ROFDMSM^{(k)}(\alpha_1, \alpha_2, \ldots, \alpha_n) =$$

$$\left(\left(\left(1 - \left(1 - \prod_{1 \le i_1 < \ldots < i_k \le n} \left(1 - \prod_{j=1}^{k} \left(1 - u_{i_j}^q \right) \right)^{1/C_n^k} \right)^{\frac{1}{k}} \right)^{\frac{1}{q}}, \left(1 - \prod_{1 \le i_1 < \ldots < i_k \le n} \left(1 - \prod_{j=1}^{k} v_{i_j}^q \right)^{1/C_n^k} \right)^{\frac{1}{qk}} \right) \right).$$

$$\tag{4}$$

Proof. We first prove the above equation is kept. Then we prove the aggregated value is a q-ROFN.

According to Definition 2, we can obtain

$$\overset{k}{\underset{j=1}{\oplus}} \alpha_{i_j} = \left(\left(1 - \prod_{j=1}^{k} \left(1 - u_{i_j}^q \right) \right)^{1/q}, \prod_{j=1}^{k} v_{i_j} \right),$$

and

$$\left(\overset{k}{\underset{j=1}{\oplus}} \alpha_{i_j} \right)^{1/C_n^k} = \left(\left(\left(1 - \prod_{j=1}^{k} \left(1 - u_{i_j}^q \right) \right)^{1/q} \right)^{1/C_n^k}, \left(1 - \left(1 - \prod_{j=1}^{k} v_{i_j}^q \right)^{1/C_n^k} \right)^{1/q} \right).$$

Further,

$$\underset{1 \leq i_1 < \ldots < i_k \leq n}{\otimes} \left(\overset{k}{\underset{j=1}{\oplus}} \alpha_{i_j} \right)^{1/C_n^k} = \left(\prod_{1 \leq i_1 < \ldots < i_k \leq n} \left(\left(1 - \prod_{j=1}^{k} \left(1 - u_{i_j}^q \right) \right) \right)^{1/q} \right)^{1/C_n^k},$$

$$\left(1 - \prod_{1 \leq i_1 < \ldots < i_k \leq n} \left(1 - \prod_{j=1}^{k} v_{i_j}^q \right) \right)^{1/C_n^k} \right)^{1/q})$$

Thus,

$$q - ROFDMSM^{(k)}(a_1, a_2, \ldots, a_n) = \frac{1}{k} \left(\underset{1 \leq i_1 < \ldots < i_k \leq n}{\otimes} \left(\overset{k}{\underset{j=1}{\oplus}} \alpha_{i_j} \right)^{1/C_n^k} \right)$$

$$\left(\left(1 - \left(1 - \prod_{1 \leq i_1 < \ldots < i_k \leq n} \left(1 - \prod_{j=1}^{k} \left(1 - u_{i_j}^q \right) \right)^{1/C_n^k} \right)^{1/k} \right)^{1/q}, \left(1 - \prod_{1 \leq i_1 < \ldots < i_k \leq n} \left(1 - \prod_{j=1}^{k} v_{i_j}^q \right)^{1/C_n^k} \right)^{1/qk} \right)$$

which proves that Eq. (4) is kept. For convenience, let

$$u = \left(1 - \left(1 - \prod_{1 \leq i_1 < \ldots < i_k \leq n} \left(1 - \prod_{j=1}^{k} \left(1 - u_{i_j}^q \right) \right)^{1/C_n^k} \right)^{1/k} \right)^{1/q}, v$$

$$= \left(1 - \prod_{1 \leq i_1 < \ldots < i_k \leq n} \left(1 - \prod_{j=1}^{k} v_{i_j}^q \right)^{1/C_n^k} \right)^{1/qk}$$

It is easy to prove that $0 \leq u, v \leq 1$. As $u_{i_j}^q + v_{i_j}^q \leq 1$, then $u_{i_j}^q \leq 1 - v_{i_j}^q$, and

$$u^q + v^q =$$

$$\left(\left(1 - \left(1 - \prod_{1 \leq i_1 < \ldots < i_k \leq n} \left(1 - \prod_{j=1}^{k} \left(1 - u_{i_j}^q \right) \right)^{1/C_n^k} \right)^{1/k} \right)^{1/q} \right)^q + \left(\left(1 - \prod_{1 \leq i_1 < \ldots < i_k \leq n} \left(1 - \prod_{j=1}^{k} v_{i_j}^q \right)^{1/C_n^k} \right)^{1/qk} \right)^q$$

$$\leq 1 - \left(\left(1 - \prod_{1 \leq i_1 < \ldots < i_k \leq n} \left(1 - \prod_{j=1}^{k} v_{i_j}^q \right)^{1/C_n^k} \right)^{1/qk} \right)^q + \left(\left(1 - \prod_{1 \leq i_1 < \ldots < i_k \leq n} \left(1 - \prod_{j=1}^{k} v_{i_j}^q \right)^{1/C_n^k} \right)^{1/qk} \right)^q = 1,$$

which proves that the aggregated result is a q-ROFN. Thus, the proof of Theorem 1 is completed.

In addition, the proposed q-ROFDMSM operator has the following properties.

Theorem 2 (Idempotency). Let $\alpha_i = (u_i, v_i)(i = 1, 2, \ldots, n)$ be a collection of q-ROFNs, if all the aggregated q-ROFNs are equal, i.e. $\alpha_i = \alpha = (u, v)$ for all i, then

$$q - ROFDMSM^{(k)}(\alpha_1, \alpha_2, \ldots, \alpha_n) = \alpha. \tag{5}$$

Theorem 3 (Monotonicity). Let $\alpha_i = (u_i, v_i)$ and $\beta_i = (s_i, t_i)(i = 1, 2, \ldots, n)$ be two collections of q-ROFNs, if $u_i \leq s_i$ and $v_i \geq t_i$ for all i, then

$$q - ROFDMSM^{(k)}(\alpha_1, \alpha_2, \ldots, \alpha_n) \leq q - ROFDMSM^{(k)}(\beta_1, \beta_2, \ldots, \beta_n). \tag{6}$$

Theorem 4 (Boundedness). Let $\alpha_i = (u_i, v_i)(i = 1, 2, \ldots, n)$ be a collection of q-ROFNs, then

$$\alpha^- \leq q - ROFDMSM^{(k)}(\alpha_1, \alpha_2, \ldots, \alpha_n) \leq \alpha^+, \tag{7}$$

where $\alpha^- = \left(\min_{i=1}^{n}(u_i), \max_{i=1}^{n}(v_i) \right)$ and $\alpha^+ = \left(\max_{i=1}^{n}(u_i), \min_{i=1}^{n}(v_i) \right)$.

The above theorems can be easily proved. In the following, we discuss some special cases of the proposed operators with respect to the parameters k and q.

Case 1. When $k = 1$, then the q-ROFDMSM operator reduces to the following

$$q - ROFDMSM^{(1)}(\alpha_1, \alpha_2, \ldots, \alpha_n) = \left(\prod_{i=1}^{n} \alpha_i \right)^{1/n}$$

$$= \left(\prod_{i=1}^{n} u_i^{1/n}, \left(1 - \prod_{i=1}^{n}(1 - v_i^q)^{1/n} \right)^{1/q} \right), \tag{8}$$

which is the q-rung orthopair fuzzy geometric operator proposed by Liu and Wang [15].

Case 2. When $k = 2$, then the q-ROFDMSM operator reduces to the following

$$q - ROFDMSM^{(2)}(\alpha_1, \alpha_2, \ldots, \alpha_n) = \frac{1}{2} \left(\prod_{\substack{i_1, i_2 = 1 \\ i_1 \neq i_2}}^{n} (\alpha_{i_1} + \alpha_{i_2})^{\frac{1}{n(n-1)}} \right) =$$

$$\left(\left(\left(1 - \left(1 - \left(\prod_{\substack{i_1 = 1, i_2 = 1 \\ i_1 \neq i_2}}^{n} \left(1 - \left(1 - u_{i_1}^q \right)\left(1 - u_{i_2}^q \right) \right) \right)^{\frac{1}{n(n-1)}} \right) \right)^{\frac{1}{2}} \right)^{\frac{1}{q}}, \left(1 - \left(\prod_{\substack{i_1 = 1, i_2 = 1 \\ i_1 \neq i_2}}^{n} \left(1 - v_{i_1}^q v_{i_2}^q \right) \right)^{\frac{1}{n(n-1)}} \right)^{\frac{1}{2q}} \right), \tag{9}$$

which is the q-rung orthopair fuzzy Bonferroni mean proposed by Liu and Liu [16].

Case 3. When $k = 3$, then the q-ROFDMSM operator reduces to the following

$$q - ROFDMSM^{(3)}(\alpha_1, \alpha_2, \ldots, \alpha_n) = \frac{1}{3}\left(\prod_{i_1,i_2,i_3=1}^{n}(\alpha_{i_1} + \alpha_{i_2} + \alpha_{i_3})\right) =$$

$$\left(\left(1 - \left(1 - \prod_{i_1,i_2,i_3=1}^{n}\left(1 - (1-u_{i_1}^q)(1-u_{i_2}^q)(1-u_{i_3}^q)\right)\right)^{\frac{1}{3}}\right)^{\frac{1}{q}}, \left(1 - \prod_{i_1,i_2,i_3=1}^{n}(1-(v_{i_1}v_{i_2}v_{i_3})^q)\right)^{\frac{1}{3}}\right),$$

$$(10)$$

which is the q-rung orthopair fuzzy generalized Bonferroni mean operator.

Case 4. When $k = n$, then the q-ROFDMSM operator reduces to the following

$$q - ROFDMSM^{(n)}(\alpha_1, \alpha_2, \ldots, \alpha_n) = \frac{1}{n}\left(\sum_{j=1}^{n}\alpha_{ij}\right)$$

$$= \left(1 - \prod_{j=1}^{n}\left(1 - u_{ij}^q\right)^{1/n}\right)^{1/q}, \left(\prod_{j=1}^{n}v_{ij}\right)^{1/n}, \quad (11)$$

which is the q-rung orthopair fuzzy arithmetic operator proposed by Liu and Wang [15]

Case 5. When $q = 2$, then the q-ROFDMSM operator reduces to the following

$$q - ROFDMSM^{(k)}(\alpha_1, \alpha_2, \ldots, \alpha_n) = \frac{1}{k}\left(\bigotimes_{1 \leq i_1 < \ldots < i_k \leq n}\left(\bigoplus_{j=1}^{k}\alpha_{ij}\right)^{1/C_n^k}\right) =$$

$$\left(\left(1 - \left(1 - \prod_{1 \leq i_1 < \ldots < i_k \leq n}\left(1 - \prod_{j=1}^{k}\left(1 - \gamma_{ij}^2\right)\right)^{1/C_n^k}\right)^{\frac{1}{k}}\right)^{\frac{1}{2}}, \left(1 - \prod_{1 \leq i_1 < \ldots < i_k \leq n}\left(1 - \prod_{j=1}^{k}\eta_{ij}^2\right)^{1/C_n^k}\right)^{\frac{1}{2}}\right),$$

$$(12)$$

which is the Pythagorean fuzzy dual Maclaurin symmetric mean operator.

Case 6. When $q = 1$, then the q-ROFDMSM operator reduces to the following

$$q - ROFDMSM^{(k)}(\alpha_1, \alpha_2, \ldots, \alpha_n) = \frac{1}{k}\left(\bigotimes_{1 \leq i_1 < \ldots < i_k \leq n}\left(\bigoplus_{j=1}^{k}\alpha_{ij}\right)^{1/C_n^k}\right) =$$

$$\left(1 - \left(1 - \prod_{1 \leq i_1 < \ldots < i_k \leq n}\left(1 - \prod_{j=1}^{k}\left(1 - \gamma_{ij}\right)\right)^{1/C_n^k}\right)^{1/k}, \left(1 - \prod_{1 \leq i_1 < \ldots < i_k \leq n}\left(1 - \prod_{j=1}^{k}\eta_{ij}\right)^{1/C_n^k}\right)^{1/k}\right),$$

$$(13)$$

which is the intuitionistic fuzzy dual Maclaurin symmetric mean operator.

3.2   The q-Rung Orthopair Fuzzy Weighted Dual Maclaurin Symmetric Mean (q-ROFWDMSM) Operator

Definition 6. Let $\alpha_i = (u_i, v_i)(i = 1, 2, \ldots, n)$ be a collection of q-ROFNs, $w = (w_1, w_2, \ldots, w_n)^T$ be the weight vector of $\alpha_i(i = 1, 2, \ldots, n)$, satisfying $w_i \in [0, 1]$ and $\sum_{i=1}^{n} w_i = 1$, and $k = 1, 2, \ldots, n$. If

$$q - ROFWDMSM^{(k)}(\alpha_1, \alpha_2, \ldots, \alpha_n) = \frac{1}{k} \left(\underset{1 \leq i_1 < \ldots < i_k \leq n}{\otimes} \left(\overset{k}{\underset{j=1}{\oplus}} w_{i_j} \alpha_{i_j} \right) \right)^{1/C_n^k}, \quad (14)$$

then $q - ROFWDMSM^{(k)}$ is called q-ROFWDMSM, where $(i_1, i_2, \ldots, i_k)$ traversal all the k-tuple combination of $(1, 2, \ldots, n)$, C_n^k is the binomial coefficient.

Similar to the q-ROFDMSM operator, the following theorem can be obtained.

Theorem 5. Let $\alpha_i = (u_i, v_i)(i = 1, 2, \ldots, n)$ be a collection of q-ROFNs, and $k = 1, 2, \ldots, n$, then the aggregated result by the q-ROFWDMSM operator is also a q-ROFN and

$$q - ROFWDMSM^{(k)}(\alpha_1, \alpha_2, \ldots, \alpha_n) = \left(\left(1 - \left(1 - \prod_{1 \leq i_1 < \ldots < i_k \leq n} \left(1 - \prod_{j=1}^{k} \left(1 - u_{i_j}^{qw_{i_j}} \right) \right)^{\frac{1}{C_n^k}} \right)^{\frac{1}{k}} \right)^{\frac{1}{q}}, \right.$$
$$\left. \left(1 - \prod_{1 \leq i_1 < \ldots < i_k \leq n} \left(1 - \prod_{j=1}^{k} \left(1 - \left(1 - v_{i_j}^{q} \right)^{w_{i_j}} \right) \right)^{\frac{1}{C_n^k}} \right)^{\frac{1}{qk}} \right),$$

$$(15)$$

Similarly, the proposed q-ROFWDMSM operator also have the properties of monotonicity and boundedness.

4   A Novel Approach for q-Rung Orthopair Fuzzy MCDM

In this section, we propose a novel approach for MCDM problems based on the proposed operators. A typical MCDM problem can be described as follows. Let $X = \{x_1, x_2, \ldots, x_m\}$ be a set of alternatives and $C = \{C_1, C_2, \ldots, C_n\}$ be a set of criteria. Weight vector of the criteria is $w = (w_1, w_2, \ldots, w_n)^T$, with the condition that $0 \leq w_j \leq 1$ and $\sum_{j=1}^{n} w_j = 1$. A decision maker is required to express his/her preference information over alternatives by a q-rung orthopair fuzzy decision matrix $A = (\alpha_{ij})_{m \times n}$, where $\alpha_{ij} = (u_{ij}, v_{ij})$ is a q-ROFN, representing the value of criteria C_j of x_i. In the following, we present a novel approach to solve the above problem on the basis of the proposed operators.

Step 1. Normalize the original decision matrix. Basically, there are two types of criteria, i.e. the benefit type and cost type. Thus, the original decision matrix should be normalized according to the following equation.

$$\alpha_{ij} = \begin{cases} (u_{ij}, v_{ij}) & C_1 \in I_1 \\ (v_{ij}, u_{ij}) & C_2 \in I_2 \end{cases}, \tag{16}$$

where I_1 and I_2 denote the benefit type criteria and cost type criteria respectively.

Step 2. Utilize the q-ROFWDMSM operator

$$\alpha_i = q - ROFWDMSM^{(k)}(\alpha_{i1}, \alpha_{i2}, \ldots, \alpha_{in}), \tag{17}$$

to aggregate the all criteria values and a series of overall values of alternatives can be obtained.

Step 3. Compute the score functions of all the alternatives.

Step 4. Rank the alternatives according to their score functions and select the optimal alternative(s).

5 Numerical Example

In this section, we apply the proposed method in a practical decision making problem to show the validity of the proposed method. To stimulate the research enthusiasm of doctoral student, Beijing Jiaotong University holds an academic forum. All doctoral student are encouraged to submit papers to the forum and make presentations. According to the regulation of the forum, the academic committee will select a best paper and give some bonus to the author(s) of the best paper. After primary evaluations, there are five papers and the best paper will be chosen from them. For convenience, let $X = \{x_1, x_2, x_3, x_4, x_5\}$ be the five papers. To select the best paper, the five alternatives are evaluated from four criteria: (1) the originality, novelty and significance of results (C_1); (2) the technical quality of work (C_2); (3) the comprehensibility and presentation of paper (C_3); (4) the overall impression (C_4). The weight vector of the criteria is $w = (0.2, 0.1, 0.3, 0.4)^T$. The committee evaluates the five possible alternatives under the above criteria by intuitionistic fuzzy information, which is listed in the following matrix (Table 1).

In the following, we utilize the proposed method to solve the above problem.

Table 1. The original intuitionistic fuzzy decision matrix

	C_1	C_2	C_3	C_4
x_1	(0.4, 0.5)	(0.5, 0.4)	(0.2, 0.7)	(0.2, 0.5)
x_2	(0.6, 0.4)	(0.6, 0.3)	(0.6, 0.3)	(0.3, 0.6)
x_3	(0.5, 0.5)	(0.4, 0.5)	(0.4, 0.4)	(0.5, 0.4)
x_4	(0.7, 0.2)	(0.5, 0.4)	(0.2, 0.5)	(0.3, 0.7)
x_5	(0.5, 0.3)	(0.3, 0.4)	(0.6, 0.2)	(0.4, 0.4)

5.1   The Decision Making Process

Step 1. As the criteria are benefit, the original decision matrix does not to be normalized.

Step 2. Utilize the q-ROFWDMSM operator to aggregate the criteria values of each alternative to calculate the overall preference values α_i of alternative $x_i(i = 1, 2, 3, 4, 5)$. Thus, we can get (let $q = 3$ and $k = 2$),

$$\alpha_1 = (0.1941, 0.8653), \ \alpha_2 = (0.296, 0.8134) \ \alpha_3 = (0.2898, 0.8249),$$
$$\alpha_2 = (0.2698, 0.8353) \ \alpha_5 = (0.3036, 0.7638)$$

Step 3. Calculate the scores $S(\alpha_i)(i = 1, 2, 3, 4, 5)$ of the overall preference values $\alpha_i(i = 1, 2, 3, 4, 5)$. We have

$$S(\alpha_1) = -0.6407, \ S(\alpha_2) = -0.5024, \ S(\alpha_3) = -0.5370, \ S(\alpha_4) = -0.5632, \ S(\alpha_5)$$
$$= -0.4175$$

Step 4. Rank the corresponding alternatives $x_i(i = 1, 2, 3, 4, 5)$ according to the rank of the overall preference values $\alpha_i(i = 1, 2, 3, 4, 5)$. We have $x_5 > x_2 > x_3 > x_4 > x_1$, i.e. x_5 is the best paper.

5.2   The Influence of the Parameters on the Results

It is noted that the proposed q-ROFWDMDM has two important parameters, i.e. q and k. These two parameters have significant influence on the scores and ranking results. We firstly assign different values to k and present the score function and raking orders in Table 2.

From Table 2, it can be seen that different score function can be obtained with different values of k in the q-ROFWDMSM operator. However, the ranking orders are

Table 2.   The score functions and ranking orders with different value of the parameter k in the q-ROFWDMSM operator ($q = 3$)

k	Score function $S(\alpha_i)(i = 1, 2, 3, 4, 5)$	Ranking orders
$k = 1$	$S(\alpha_1) = -0.6584 \ S(\alpha_2) = -0.5254 \ S(\alpha_3) = -0.5862$ $S(\alpha_4) = -0.5933 \ S(\alpha_5) = -0.4798$	$x_5 > x_2 > x_3 > x_4 > x_1$
$k = 2$	$S(\alpha_1) = -0.6407 \ S(\alpha_2) = -0.5024 \ S(\alpha_3) =$ $-0.5370 \ S(\alpha_4) = -0.5632 \ S(\alpha_5) = -0.4175$	$x_5 > x_2 > x_3 > x_4 > x_1$
$k = 3$	$S(\alpha_1) = -0.6306 \ S(\alpha_2) = -0.4913 \ S(\alpha_3) = -0.5160$ $S(\alpha_4) = -0.5480 \ S(\alpha_5) = -0.3956$	$x_5 > x_2 > x_3 > x_4 > x_1$
$k = 4$	$S(\alpha_1) = -0.6227 \ S(\alpha_2) = -0.4826 \ S(\alpha_3) = -0.5026$ $S(\alpha_4) = -0.5365 \ S(\alpha_5) = -0.3805$	$x_5 > x_2 > x_3 > x_4 > x_1$

always $x_5 > x_2 > x_3 > x_4 > x_1$. In addition, it can be found that with the increase of the value of k in the q-ROFWDMSM operator, the score function of the overall preference values of alternatives also increase. Thus, the value of k can be viewed as decision makers' attitude to optimism and pessimism in practical decision making problems. If decision makers are optimistic to their decisions, then the greater value should be assigned to k, whereas if decision makers are pessimistic to their decisions, the less values should be assigned to k. This characteristic illustrates the flexibility and powerfulness of the proposed method.

In the following, we investigate the effects of q on the score functions and ranking results. We assign different values to q in the q-ROFWDMSM operator and represent the score functions and ranking orders in Figs. 1 and 2. It is noted that we investigate the influence of the parameter q in the q-ROFWDMSM operator when $k = 2$ and 3. This is because when $k = 1$ and 4, the q-ROFWDMSM operator reduces to the weighted averaging and geometric operators respectively, which do not consider the interrelationship among attributes. In other words, we only consider cases in which the interrelationship among attributes is considered. It can be seen from the Figs. 1 and 2, with the increase of q in the q-ROFWDMSM operator, the score function of overall preference values also increase, whereas the ranking orders are always the same. This also demonstrates the flexibility and powerfulness of the proposed method.

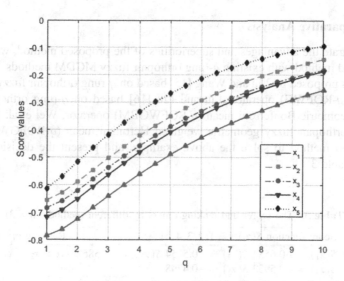

Fig. 1. The score function and ranking orders by the q-ROFWDMSM operator with different values of q ($k = 2$)

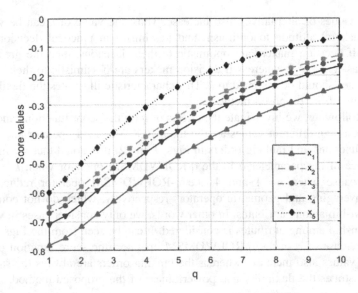

Fig. 2. The score function and ranking orders by the q-ROFWDMSM operator with different values of q ($k = 3$)

5.3 Comparative Analysis

To demonstrate the advantages and superiorities of the proposed method, we compare the proposed method with existing q-rung orthopair fuzzy MCDM methods. We utilize the methods proposed by Liu and Wang [15] based on q-rung orthopair fuzzy weighted geometric (q-ROFWG) operator, Liu and Liu [16] based on q-rung orthopair fuzzy weighted geometric Bonferroni mean (q-ROFWGBM) operator, Wei et al. [18] based on q-rung orthopair fuzzy geometric weighted Heronian mean (q-ROFGWHM), and the proposed method to solve the above example, and present the decision making results in Table 3.

Table 3. The scores and ranking orders by different methods ($q = 3$)

Method	Score function $S(\alpha_i)(i = 1, 2, 3, 4, 5)$	Ranking orders
Method in [15]	$S(\alpha_1) = -0.6584$ $S(\alpha_2) = -0.5254$ $S(\alpha_3) = -0.5862$ $S(\alpha_4) = -0.5933$ $S(\alpha_5) = -0.4798$	$x_5 > x_2 > x_3 > x_4 > x_1$
Method in [16] ($s = t = 1$)	$S(\alpha_1) = -0.6331$ $S(\alpha_2) = -0.3916$ $S(\alpha_3) = -0.5963$ $S(\alpha_4) = -0.6078$ $S(\alpha_5) = -0.4007$	$x_2 > x_5 > x_3 > x_4 > x_1$
Method in [18] ($s = t = 1$)	$S(\alpha_1) = -0.6318$ $S(\alpha_2) = -0.4001$ $S(\alpha_3) = -0.5937$ $S(\alpha_4) = -0.6126$ $S(\alpha_5) = -0.4238$	$x_2 > x_5 > x_3 > x_4 > x_1$
Method in this paper ($k = 3$)	$S(\alpha_1) = -0.6306$ $S(\alpha_2) = -0.4913$ $S(\alpha_3) = -0.5160$ $S(\alpha_4) = -0.5480$ $S(\alpha_5) = -0.3956$	$x_5 > x_2 > x_3 > x_4 > x_1$

From Table 3, we can find out that the scores of the alternatives derived by Liu and Wang's [15] method are the same as the scores obtained by the proposed method in this paper when $k = 1$. This is because Liu and Wang's [15] method is a special case of the proposed method. Thus, our method is more general than Liu and Wang's [15] method. Moreover, our method can reflect the interrelationship among criteria values, whereas Liu and Wang's [15] method cannot. The methods proposed by Liu and Liu [16] and Wei et al. [18] can also reflect the interrelationship between arguments. However, the results derived by Liu and Liu's [16] and Wei et al.'s [18] methods are still different from the result derived by the proposed method. This is because the proposed method in this paper can capture the interrelationship among multiple input arguments, whereas Liu and Liu' [16] and Wei et al.'s [18] methods can only reflect the interrelationship between any two criterion values. In the best paper selection problem, interrelationship exist among multiple criteria values. Thus, the proposed method is more suitable than the methods proposed Liu and Liu [16], and Wei et al. [18] for addressing such problem.

6 Conclusions

The recently developed q-ROFSs are a good tool to cope with fuzzy decision information in the process of MCDM. Recently, some q-rung orthopair fuzzy aggregation operators for fusing q-rung orthopair fuzzy information have been developed. Nevertheless, these operators fail to capture the interrelationship among multiple criterion values on real-life MCDM problems. Motivated by the DMSM, this paper proposes some q-rung orthopair fuzzy MSM operator. Further, a novel approach for q-rung orthopair fuzzy MCDM is developed. A best paper selection example has proved the validity and merits of the proposed method. In the future, we will apply the proposed method in actual decision making problems.

Acknowledgements. This work was partially supported by a key program of the National Natural Science Foundation of China (NSFC) with grant number 71532002 and the Fundamental Research Funds for the Central Universities with grant number 2017YJS075.

References

1. Atanassov, K.T.: Intuitionistic fuzzy sets. Fuzzy Sets Syst. **20**(1), 87–96 (1986)
2. Zadeh, L.A.: Fuzzy sets. Inform. Contr. **8**, 338–356 (1965)
3. Chen, S.M., Chang, C.H.: Fuzzy multiattribute decision making based on transformation techniques of intuitionistic fuzzy values and intuitionistic fuzzy geometric averaging operators. Inform. Sci. **352**, 133–149 (2016)
4. Xu, Z.S.: Some similarity measures of intuitionistic fuzzy sets and their applications to multiple attribute decision making. Fuzzy Optim. Decis. Ma. **6**(2), 109–121 (2007)
5. Zhao, Q.Y., Chen, H.Y., Zhou, L.G.: The properties of fuzzy number intuitionistic fuzzy prioritized operators and their applications to multi-criteria group decision making. J. Intell. Fuzzy Syst. **28**(4), 1835–1848 (2015)

6. Xu, Z.S.: A method based on distance measure for interval-valued intuitionistic fuzzy group decision making. Inform. Sci. **180**, 181–190 (2010)
7. Yager, R.R.: Pythagorean membership grades in multicriteria decision making. IEEE Tran. Fuzzy Syst. **22**(4), 958–965 (2014)
8. Zhang, X.L., Xu, Z.S.: Extension of TOPSIS to multiple criteria decision making with Pythagorean fuzzy sets. Int. J. Intell. Syst. **29**(12), 1061–1078 (2014)
9. Liang, D.C., Xu, Z.S., Liu, D., Wu, Y.: Method for three-way decisions using ideal TOPSIS solutions at Pythagorean fuzzy information. Inform. Sci. **435**, 282–295 (2018)
10. Perez-Dominguez, L., Rodriguez-Picon, L.A., Alvarado-Iniesta, A., Cruz, D.L., Xu, Z.S.: MOORA under Pythagorean fuzzy set for multiplie criteria decision making. Complexity. Article ID 2602376
11. Xing, Y.P., Zhang, R.T., Wang, J., Zhu, X.M.: Some new Pythagorean fuzzy Choquet–Frank aggregation operators for multi-attribute decision making. Int. J. Intell. Syst. (2018, Published online)
12. Zhang, R.T., Wang, J., Zhu, X.M., Xia, M.M., Yu, M.: Some generalized Pythagorean fuzzy Bonferroni mean aggregation operators with their application to multiattribute group decision-making. Complexity. Article ID 5937376
13. Xu, Y., Shang, X.P., Wang, J.: Pythagorean fuzzy interaction Muirhead means with their application to multi-attribute group decision-making. Information **9**(7), 157 (2018)
14. Yager, R.R.: Generalized orthopair fuzzy sets. IEEE Tran. Fuzzy Syst. **25**(5), 1222–1230 (2017)
15. Liu, P.D., Wang, P.: Some q-rung orthopair fuzzy aggregation operators and their applications to multiple-attribute decision making. Int. J. Intell. Syst. **33**(2), 259–280 (2018)
16. Liu, P.D., Liu, J.L.: Some q-rung orthopair fuzzy Bonferroni mean operators and their application to multi-attribute group decision making. Int. J. Intell. Syst. **33**(2), 315–347 (2018)
17. Liu, P.D., Wang, P.: Multiple-attribute decision making based on Archimedean Bonferroni operators of q-rung orthopair fuzzy numbers. IEEE Tran. Fuzzy Syst. (2018, Online)
18. Wei, G.W., Gao, H., Wei, Y.: Some q-rung orthopair fuzzy Heronian mean operators in multiple attribute decision making. Int. J. Intell. Syst. **33**(7), 1426–1458 (2018)
19. Maclaurin C.: A second letter to martin folkes, esq.; concerning the roots of equations, with demonstration of other rules of algebra. Philos. Trans. R Soc. Lon. Ser. A **36**, 59–96 (1729)
20. Qin, J.D., Liu, X.: Approaches to uncertain linguistic multiple attribute decision making based on dual Maclaurin symmetric mean. J. Intell. Fuzzy Syst. **29**(1), 171–186 (2015)

An Improved Short Pause Based Voice Activity Detection Using Long Short-Term Memory Recurrent Neural Network

Kiettiphong Manovisut[1(✉)], Pokpong Songmuang[1],
and Nattanun Thatphithakkul[2]

[1] Department of Computer Science, Faculty of Science and Technology,
Thammasat University, Pathum Thani, Thailand
`manovisut.ktp@gmail.com, pokpong@cs.tu.ac.th`
[2] National Electronics and Computer Technology Center, Pathum Thani, Thailand
`nattanun.thatphithakkul@nectec.or.th`

Abstract. Generally, voice activity detection (VAD) commonly uses a silence over 100-ms as an endpoint of speech. Previously, the short pause based VAD is proposed to reduce the waiting time of caption result in automatic captioned relay service. This technique reduces the waiting time of caption result well. However, an accuracy of caption result is not maintained as it should be. The problem inherits to short-time energy feature which difficult and inaccurate to search the smallest characteristic like short pause or unvoiced sounds. Therefore, we propose the new technique that combines a Mel Frequency Cepstral Coefficient and Long Short-term Memory Recurrent Neural Network. This technique is called a pause classifier, which is able to capture the smallest characteristic like the short pause or unvoiced sounds. The experimental result shows an effective to reduce the waiting time while maintaining WER of caption result. The average waiting time reduced, the automatic speech recognition results are more continuous and constant. This will directly affect the user experience in automatic captioned relay service.

Keywords: Short pause · Voice activity detection
Long short-term memory · Recurrent neural network

1 Introduction

Voice activity detection (VAD) is a speech processing technique that detects the presence or absence of human speech. VAD is an essential technology for a variety of speech-based applications such as speech coding, speech recognition, and speech enhancement. Therefore, various VAD algorithms provide varying features and compromise between latency, sensitivity, accuracy and computational cost differently.

© Springer Nature Singapore Pte Ltd. 2018
J. Chen et al. (Eds.): KSS 2018, CCIS 949, pp. 267–274, 2018.
https://doi.org/10.1007/978-981-13-3149-7_20

Previously, traditional short pause based VAD [6] is proposed to reduce the waiting time of caption result in automatic captioned relay service. Instead of using only silence at 100-ms, the traditional short pause based VAD uses a short pause range from 80-ms as an endpoint of speech. This technique reduces the waiting time of caption.

Since we use the short pause instead of using only silence, there is an opportunity to dismiss the unvoiced sounds such as F, P, S, T, and Th sound. These types of speech are statistically similar to background noise. This problem directly affects the accuracy of caption result due to a fault of short pause and unvoiced detection. The problem inherits to short-time energy feature which determines speech portion using threshold energy. A characteristic of short pause and unvoiced sound are the smallest characteristic that can occur in speech. Hence, using short-time energy will be difficult and inaccurate to capture the smallest characteristics.

The goal of this paper is how to accurate the short pause and unvoiced sound detection in traditional short pause based VAD. In this paper, we propose a pause classifier that combines a Mel Frequency Cepstral Coefficient (MFCC) and Long Short-term Memory (LSTM) Recurrent Neural Network (RNN) to capture the smallest characteristic like the short pause or unvoiced sounds. MFCC feature extraction is used to break apart complex sound wave into frequency bands. Afterward, we use LSTM to discriminate the MFCC features into a speech or pause in a frame by frame. These techniques improve the accuracy of short pause and unvoiced detection.

In this paper, the background knowledge of traditional short pause based VAD, MFCC, and LSTM are presented in the Sect. 2. Section 3, the creation of the pause classifier has been discussed. Next, the experiment of the proposed work is shown in Sect. 4. Finally, the experiment results and discussion are described in Sect. 5.

2   Background Knowledge

2.1   Short Pause Based VAD

Generally, VADs commonly use a silence over 100-ms as an endpoint of speech for better accuracy [7, 10]. Traditional short pause based VAD aims to reduce the waiting time of automatic speech recognition (ASR) result. The VAD provides short pause algorithm which use short pauses in speech range from 80-ms instead using only silence at 100-ms as an endpoint. Moreover, the padding silence and endpoint decision are used to maintain the accuracy of result caused by using short pause. These techniques reduce the waiting time and maintain the accuracy of result well.

Previously, traditional short pause based VAD is improved from traditional dual-threshold method which uses short-time energy and zero-crossing rate as the feature extractions. However, an accuracy of caption result is not maintained as it should be. The problem inherits to short-time energy feature which determines speech portion using threshold energy. The short-time energy cannot capture the smallest characteristic like the short pause or unvoiced sounds. This problem always related to an accuracy of result directly.

Hence, we propose a new technique that combines a Mel Frequency Cepstral Coefficient (MFCC) and Long Short-term Memory (LSTM) Recurrent Neural Network (RNN). This technique is called a pause classifier, which is able to capture the smallest characteristic like the short pause or unvoiced sound during speech. The description of MFCC and LSTM are described in the next topic.

2.2 Mel Frequency Cepstral Coefficient (MFCC)

MFCC is a feature widely used in ASR, especially with Hidden Markov Model (HMM) classifiers. MFCC is used to extract a feature vector containing all information about parts of the human speech production and speech perception. To represent the dynamic nature of speech the MFCC also includes the change of the feature vector over time as part of the feature vector.

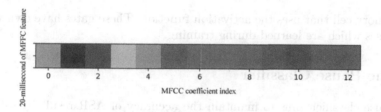

Fig. 1. An example of MFCC coefficient index

The Fig. 1 is an example of 13-MFCC coefficients. Each MFCC coefficient index represents how much energy of each frequency band in speech frame.

2.3 Long Short-Term Memory Recurrent Neural Network

In the recent years, neural networks become popular in many tasks such as image processing, natural language processing and speech classification. Long short-term memory network (LSTM) [3] is a special kind of recurrent neural network (RNN) that enables the networks to process and learn data sequences. LSTM is mainly designed to avoid the long-term dependency problem. LSTM is widely used in speech classification for improving speech and non-speech classifier. Many proposed works [2,5,8] show that LSTM improves an accuracy and noise robustness using the context of previous speech frames well.

The figure of LSTM unit is illustrated in Fig. 2. LSTM unit contains three gates in memory cell including an input gate, forget gate and output gate. c_t is a cell state, h_t is a hidden state and x_t is an input at time t. The input gate decides the new information into the memory cell using x_t and h_t. The forget gate decides the value that will keep in the memory cell using the previous hidden state (h_{t-1}) and previous cell state (c_{t-1}). The output gate decides the value in

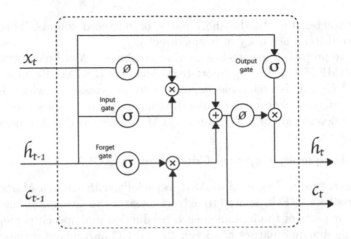

Fig. 2. An illustrated of LSTM unit

the memory cell that uses the activation function. These gates have own weight and biases which are learned during training.

3   The Pause Classifier

The pause classifier aims to maintain the accuracy of ASR result by solving a weakness of the short pause and unvoiced detection. The pause classifier combines MFCC features and LSTM to capture the smallest characteristic like the short pause and unvoiced sounds. To create the classifier, we slice speech audio into 20-ms of chunk. Then MFCC feature extraction is used to break apart complex sound wave into frequency bands. Afterwards, we use LSTM to classify the MFCC features into a speech or pause in frame by frame. The output of the pause classifier is binary classification (0 to 1) that represents a probability of pause or speech.

We replace the short-time energy with the pause classifier. Since the pause classifier is learned from speech and unvoiced speech, an unvoiced searching by zero-crossing rate is no longer used in this paper. The flow diagram of proposed work is illustrated in the Fig. 3.

We use 12 h of speech sentence from LOTUS [4], which is labeled short pause and silence by hand. The hand labels and speech audios are used to create the pause classifier. The dataset is separated to 80% for training and then test a result with remaining 20%.

4   Experiment

In the experiment, 20 min of LOTUS-BN dataset [1] is used to measure an average waiting time of ASR result. LOTUS-BN is a good resource for investigating on the waiting time of ASR. Since the LOTUS-BN dataset is a Thai television

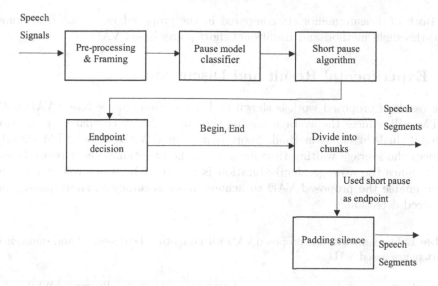

Fig. 3. The flow diagram of improved short pause based VAD with LSTM

broadcast news corpus, it has a higher rate of speech (approximately 196 words per minute) more than natural speech (150 words per minute) [9]. This is an ideal for analyzing the waiting time caused by silence or endpoint detection in VAD. An audio signal is 16 kHz sampling rate and encoded in a 16 bits Microsoft PCM format. Please note that the sample rate of speech audio is reduced to 8 kHz in this experiment.

We study a minimum short pause in different length that suitable for the algorithm. The minimum short pauses are 20, 40, 60 and 80 ms, respectively (range of the short pause caused by framing at 20-ms).

An average waiting time and word error rate (WER) are used for the measurement. The average waiting time for ASR result is included the detection time of VAD, upload time and recognition time. The equation of average waiting time is as follows:

$$AVG_{wt} = \frac{\sum_{i=0}^{N-1} D_i + (U_i) + (Ci \times RTF)}{N-1} \qquad (1)$$

where a detection time of VAD is D_i. U_i represents an upload time for i chunk. C_i is a duration time of speech chunk. N is a number of ASR result. Finally, RTF is a processing time factor of ASR (also known as a real-time factor). Due to a large number of ASR providers, we set 1× RTF to control external factor that affects the experiment.

WER is a number of substitution, deletion and insertion error over a number of correct words as the following equation:

$$WER = \frac{Substitution + Deletion + Insertion}{Substitution + Deletion + Correct} \qquad (2)$$

Both of measurements are compared in the proposed work, the traditional dual-threshold method and traditional short pause based VAD.

5   Experimental Result and Discussion

The result of proposed work is shown in Table 1. Short pause based VAD with LSTM still reduces the average waiting time up to 17.1% compared to the traditional dual-threshold method. Short pause based VAD with LSTM slightly reduces the average waiting time lower than the traditional short pause based VAD. Since the average chunk duration is increased by an unvoiced portion, these enable the proposed VAD to achieve more accuracy in short pause and unvoiced detection.

Table 1. The comparison of proposed VAD with dual-threshold method and traditional short pause based VAD

Algorithms	Average chunk duration (s)	Average waiting time (s)	Reduced waiting time (%)	WER (%)
Traditional Dual-threshold Method	2.14	4.70	-	26.7
Traditional Short Pause based VAD	1.68	3.71	21.2	27.6
Short Pause based VAD with LSTM	**1.77**	**3.90**	**17.1**	**25.3**

Additionally, short pause based VAD with LSTM reduces WER of result up to 1.5% and 2.29% compared with other VADs. The average waiting time of proposed VAD is reduced from 4.7 to 3.9 s at least.

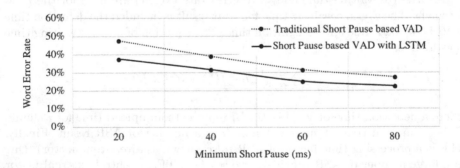

Fig. 4. The comparison of WER in traditional short pause based VAD and short pause based VAD with LSTM

Although, the average waiting time is not reduced better than traditional short pause based VAD which is reduced the average waiting time up to 21.2%.

However, the short pause based VAD with LSTM reduces WER to 25.3%, which is better than traditional dual-threshold method and traditional short pause based VAD. The comparison of WER in Fig. 4 shows that the short pause based VAD with LSTM maintains the accuracy and efficiency of caption result well, which is suitable for use in automatic captioned relay service.

6 Conclusion

Previously, the traditional short pause based VAD is proposed to reduce the waiting time of caption result in automatic captioned relay service. In this paper, we propose a technique that combines MFCC and LSTM to improve the accuracy of the caption result by improving short pause and unvoiced sound detection. The experimental result shows the efficiency to reduce the waiting time 17.1% at least. Moreover, the WER of the proposed VAD is reduced to 25.3%, which is the lowest WER compared with the traditional dual-threshold method and traditional short pause based VAD. Reducing average waiting time of caption result, the caption are more continuous and constant. This will directly affect the user experience in automatic captioned relay service.

References

1. Chotimongkol, A., Saykhum, K., Chootrakool, P., Thatphithakkul, N., Wutiwi-watchai, C.: LOTUS-BN: a Thai broadcast news corpus and its research applications. In: 2009 Oriental COCOSDA International Conference on Speech Database and Assessments ICSDA 2009, pp. 44–50 (2009). https://doi.org/10.1109/ICSDA.2009.5278377
2. Eyben, F., Weninger, F., Squartini, S., Schuller, B.: Real-life voice activity detection with LSTM recurrent neural networks and an application to Hollywood movies. In: 1988 International Conference on Acoustics, Speech, and Signal Processing ICASSP-1988, pp. 483–487. IEEE Signal Processing Society, Institute of Electrical and Electronics Engineers (2013). https://doi.org/10.1109/ICASSP.2013.6637694
3. Hochreiter, S., Schmidhuber, J.: Long short-term memory. Neural Comput. **9**(8), 1735–1780 (1997). https://doi.org/10.1162/neco.1997.9.8.1735
4. Kasuriya, S., Sornlertlamvanich, V., Cotsomrong, P., Kanokphara, S., Thatphithakkul, N.: Thai speech corpus for Thai speech recognition. In: Proceedings of Oriental COCOSDA, pp. 54–61, January 2003
5. Kim, J., Kim, J., Lee, S., Park, J., Hahn, M.: Vowel based voice activity detection with LSTM recurrent neural network. In: Proceedings of the 8th International Conference on Signal Processing Systems - ICSPS 2016, pp. 134–137 (2016). https://doi.org/10.1145/3015166.3015207
6. Manovisut, K.: Reducing waiting time in automatic captioned relay service using short pause in voice activity detection. In: 2017 9th International Conference on Knowledge and Smart Technology (KST), pp. 216–219 (2017)
7. Moattar, M., Homayounpour, M.: A simple but efficient real-time voice activity detection algorithm. In: European Signal Processing Conference (EUSIPCO) (Eusipco), pp. 2549–2553 (2009). https://doi.org/10.1007/978-1-4419-1754-6

8. Sertsi, P., Boonkla, S., Chunwijitra, V., Kurpukdee, N., Wutiwiwatchai, C.: Robust voice activity detection based on LSTM recurrent neural networks and modulation spectrum. In: 2017 Asia-Pacific Signal and Information Processing Association Annual Summit and Conference (APSIPA ASC), pp. 342–346, December 2017. http://ieeexplore.ieee.org/document/8282048/, https://doi.org/10.1109/APSIPA.2017.8282048

9. The National Center for Voice and Speech (NCVS): Voice Qualities (2007). http://www.ncvs.org/ncvs/tutorials/voiceprod/tutorial/quality.html

10. Wu, S.L., Kingsbury, B.E., Morgan, N., Greenberg, S.: Incorporating information from syllable-length time scales into automatic speech recognition. In: International Conference on Acoustics, Speech and Signal Processing (ICASSP), vol. 2, pp. 721–724 (1998). https://doi.org/10.1109/ICASSP.1998.675366

Author Index

Printed in the United States
By Bookmasters